Shrinking Cities

Shrinking Cities offers a contemporary look at patterns of shrinkage and decline in the United States. The book juxtaposes the complex and numerous processes that contribute to these patterns with broader policy frameworks that have been under consideration to address shrinkage in U.S. cities. A range of methods is employed to answer theoretically grounded questions about patterns of shrinkage and decline, the relationships between the two, and the empirical associations among shrinkage, decline, and several socio-economic variables. In doing so, the book examines new spaces of shrinkage in the United States. The book also explores pro-growth and decline-centered governance, which has important implications for questions of sustainability and resilience in U.S. cities. Finally, the book draws attention to U.S.-wide demographic shifts and argues for further research on socio-economic pathways of various groups of population, contextualized within population trends at various geographic scales. This timely contribution contends that an understanding of what the city has become, as it faces shrinkage, is essential toward a critical analysis of development both within and beyond city boundaries. The book will appeal to urban and regional studies scholars from a variety of disciplinary backgrounds, as well as practitioners and policymakers.

Russell Weaver is an Assistant Professor in the Department of Geography at Texas State University, USA

Sharmistha Bagchi-Sen is a Professor of Geography at the State University of New York at Buffalo, USA

Jason Knight is an Assistant Professor in the Department of Geography and Planning at the State University of New York at Buffalo, USA

Amy E. Frazier is an Assistant Professor of Geography at Oklahoma State University, USA.

Routledge Studies in Urbanism and the City

This series offers a forum for original and innovative research that engages with key debates and concepts in the field. Titles within the series range from empirical investigations to theoretical engagements, offering international perspectives and multidisciplinary dialogues across the social sciences and humanities, from urban studies, planning, geography, geohumanities, sociology, politics, the arts, cultural studies, philosophy and literature.

For a full list of titles in this series, please visit www.routledge.com/series/RSUC

The Urban Political Economy and Ecology of Automobility
Driving cities, driving inequality, driving politics
Edited by Alan Walks

Cities and Inequalities in a Global and Neoliberal World
Edited by Faranak Miraftab, David Wilson and Ken E. Salo

Beyond the Networked City
Infrastructure reconfigurations and urban change in the North and South
Edited by Olivier Coutard and Jonathan Rutherford

Technologies for Sustainable Urban Design and Bioregionalist Regeneration
Dora Francese

Markets, Places, Cities
Kirsten Seale

Shrinking Cities
Understanding urban decline in the United States
Russell C. Weaver, Sharmistha Bagchi-Sen, Jason C. Knight, and Amy E. Frazier

Mega-Urbanization in the Global South
Fast cities and new urban utopias of the postcolonial state
Edited by Ayona Datta and Abdul Shaban

Shrinking Cities

Understanding urban decline in the United States

Russell Weaver, Sharmistha Bagchi-Sen, Jason Knight, and Amy E. Frazier

Routledge
Taylor & Francis Group
LONDON AND NEW YORK

First published 2017
by Routledge
2 Park Square, Milton Park, Abingdon, Oxon OX14 4RN

and by Routledge
711 Third Avenue, New York, NY 10017

Routledge is an imprint of the Taylor & Francis Group, an informa business

British Library Cataloguing in Publication Data
A catalogue record for this book is available from the British Library

Library of Congress Cataloging-in-Publication Data
Names: Weaver, Russell, author.
Title: Shrinking cities : understanding urban decline in the United States / Russell Weaver, Sharmistha Bagchi-Sen, Jason Knight and Amy E. Frazier.
Description: Abingdon, Oxon ; New York, NY : Routledge, 2017. | Includes bibliographical references and index.
Identifiers: LCCN 2016011423 | ISBN 9781138796867 (hardback) | ISBN 9781315757582 (ebook)
Subjects: LCSH: Shrinking cities—United States. | Cities and towns—United States—Growth. | Urban policy—United States. | Urban renewal—United States. | Sociology, Urban—United States.
Classification: LCC HT384.U5 W44 2017 | DDC 307.760973—dc23
LC record available at https://lccn.loc.gov/2016011423

ISBN: 978-1-138-79686-7 (hbk)
ISBN: 978-1-315-75758-2 (ebk)

Typeset in Times New Roman
by Apex CoVantage, LLC

Contents

Figures

Tables

Boxes

Acknowledgments

Russell Weaver thanks his infinitely patient and supportive wife, Michele, and their son, Nicholas, for all of their (many) sacrifices that advanced the writing of this book. Sharmistha Bagchi-Sen thanks her daughters, Leena and Shohini, for their unconditional support. She also thanks her graduate assistants Torsten Schunder, Misa Yasumishi, and Marissa Bell. Jason Knight could not have done this book without the patient and unwavering support of Bree and Molly. He also thanks his student assistant Paula Jones. Amy Frazier would like to thank her family for their patience and support. The book would not have been possible without the individual dedication of all four authors and their collaborative spirit. Sharmistha would also like to thank the Baldy Center at the University at Buffalo for partial support of her research for the book.

1 Introduction

For decades, places around the world, particularly in many developing countries, have been experiencing intense urban development, rapid expansion of infrastructure, and massive population and economic growth. At the same time, many other cities, especially in developed countries, have been experiencing a different situation. Shrinking cities have become a widespread phenomenon, and they are now found on every continent around the world (with the exception of Antarctica). While the root causes of population loss vary from city to city (e.g., deindustrialization, suburbanization, demographic shifts, etc.), the effects of that population loss are often quite similar and manifest through vacant and abandoned buildings, loss of municipal services, and a struggling economy.

Consider two very different American cities: Detroit, Michigan, and Phoenix, Arizona. In 1900, Detroit was burgeoning into a thriving hub of manufacturing and shipping industries, its strategic position along the Great Lakes made it pivotal for shipping and global commerce. In 1900, it was the thirteenth largest city in the United States in terms of population, with 285,704 residents and growing. At this point, Detroit was on the cusp of massive population and economic growth centered on the nascent automobile industry, and by 1920 it had jumped to the fourth largest city in the country with a population of 993,079. Indeed, Detroit experienced the industrial boom that swept through the Northeast region of the country at that time, capitalizing on its optimal location for shipping, and expanding its boundaries to encompass the thousands of people migrating to the city to take advantage of the economic opportunity.

Meanwhile, seventeen hundred miles southwest of Detroit, Phoenix, Arizona was a tiny agricultural and resource extraction outpost of 5,544 residents. Heavily dependent on large-scale irrigation to grow crops, Phoenix was limited in its ability to expand both economically and in terms of population. In 1900, it did not even rank among the top 100 most populous cities in the United States. Phoenix grew slowly through the first part of the twentieth century, and by 1920, its population had barely eclipsed 29,000 residents.

Fast forward to 2014. Detroit is in crisis. The spatial reorganization of the auto industry and population out-migration has impacted the city for decades. Since its population peak in 1950, out-migration has left only 680,250 residents in the city, dropping it down to only the eighteenth largest city in the United States. With a declining number of residents, the city is unable to fund itself through taxes. There is little money to support the government. Eventually, the city files for bankruptcy. Meanwhile, seventeen hundred miles to the southwest, Phoenix has rocketed up the urban hierarchy with a diverse economy based not only on agriculture now but also on strong finance, real estate, and technology sectors. People have flocked to Phoenix to take advantage of the economic opportunities and mild climate, pushing Phoenix up to the sixth most populous city in the country with an estimated 1,537,058 residents.

On the surface, this example may appear to be just a tale of two cities, each with unique extenuating circumstances that disconnect them both from the overall story of the American city. Beneath the surface, however, the divergent histories of Detroit and Phoenix are representative of significant economic and social changes taking place in urban spaces across the United States, in particular throughout the second half of the twentieth century and continuing into the twenty-first century. The foregoing example is a story of the decline of once mighty economic powerhouse cities in the industrial region (often referred to as the Rust Belt) stretching from the Northeast through the Great Lakes and into the Midwest and the growth of cities in the southern and western United States (often referred to as the Sun Belt), where people are attracted to the warm climate and mild winters. Detroit is only one of many cities experiencing decline. Former industrial powerhouses such as Buffalo, New York; Cleveland, Ohio; and Pittsburgh, Pennsylvania, have also suffered debilitating economic and population losses as well as smaller but once-thriving places such as Gary, Indiana, and Scranton, Pennsylvania. Cities outside the Rust Belt are not immune to shrinkage either. St. Louis, Missouri, has suffered a similar fate as Detroit, Buffalo, and Pittsburgh, while New Orleans, Louisiana, faced a rapidly declining population in the wake of Hurricane Katrina.

Despite the acute cause of shrinkage, the challenges facing shrinking cities are extensive and complex. Population loss is often only one of many dimensions of shrinkage. As people migrate out of a city in search of jobs or a better quality of life, the city loses a vital tax revenue. Without a stable tax base, cities often cannot maintain critical services (e.g., police patrols, garbage pickup, etc.) across the entire extent of the city and are forced to cancel or reduce utilities. A byproduct of reducing services is the elimination of even more jobs from the area – furthering even more out-migration in some cases. A decreasing tax base also means that the funds needed to combat decline through economic incentives to stabilize or re-grow the city are not available. Shrinking cities have tried myriad approaches to "plugging the dam" including "pro-growth" agendas that support policies for investment, population, and job growth as well as "smart decline" strategies that seek to manage the loss of population, industry, taxes, services, and many other factors in a manner that supports those residents that have elected to remain in the city. We acknowledge the complexity involved with understanding shrinking cities, and in this book we offer both empirical analyses of the patterns of shrinkage and decline observed in the United States as well as discussion of the underlying processes and policies.

In cities, when a particular change affects one part of the system at a given point in time (e.g., a specific industry closes during a recession), a cascade of effects, which may be orders of magnitude greater than the initial change, may ensue (Martinez-Fernandez et al. 2012). This observation follows from Nobel Prize-winning economist Gunnar Myrdal's (1957) concept of cumulative causation (Emery and Flora 2006; Hospers 2014). Generally speaking, it means that once a change originates in a given area, inter-connections between the changed variable(s) and other local variables give rise to self-reinforcing feedback processes. Prior to World War II, these feedback processes almost unanimously pointed to a virtuous cycle of urbanization. Simply put, cities grew. Throughout the world, prewar industrialized cities enjoyed steady, positive inflows of people, jobs, income, and built structures. Indeed, the field of urban planning emerged around the same time largely from the need to control and manage these widespread, seemingly unabating patterns of city growth (Hollander et al. 2009). Urban growth did not cease after World War II – in fact, the urban share of global population has increased in every decade since 1940 (United Nations 2014) – however, by 1950 the phenomenon became far more narrow in its geographical scope. That is, whereas prewar urbanization was mostly distributive, in that it affected virtually all cities, postwar urbanization has been comparably parasitic, fueling growth in some cities while contributing to stagnation, shrinkage, and/or decline in others, especially in the United States (Beauregard 2006).

Acknowledging that there is no universally accepted definition of the concept of shrinkage (Pallagst, Wiechmann, and Martinez-Fernandez 2013), this book interprets the concept to reflect sustained, downward, quantitative adjustments to the population of a given geographic community (e.g., Schilling and Logan 2008). Stated another way, urban shrinkage involves long-term, "persistent" decreases in the total number of people who dwell in an affected, shrinking area (Beauregard 2009). Frequently, this sort of sustained population loss is accompanied, either before or after, by downward quantitative adjustments (i.e., shrinkage) in the size of the economy and built environment of the depopulating community. In other words, population loss tends to go hand in hand with both job loss and property abandonment (which is often followed by property demolition) in a shrinking place. While the precise chain of causality involved in this complex relationship remains unresolved (Großmann et al. 2013), most researchers and practitioners agree that these linkages between population, the economy, and the built environment are characterized by cumulative causation (Hospers 2014).

On that backdrop, unlike the pre-WWII era characterized by distributive growth, the contemporary era of parasitic urbanization has placed countless shrinking communities across the world into a downward spiral (negative cumulative causation) of demographic, economic, and physical change. As American urban scholar Lewis Mumford (1961: 486) observed in his book *The City in History*, decreases in the population, economy, and physical structures of a city often lead to a "breakup of the old urban form" (as cited in Beauregard 2006: 41). As a result, parasitic urbanization tends to have deleterious effects on a shrinking city's existing urban functions, putting pressures on policymakers and other stakeholders.

While the quantitative adjustments (e.g., population loss) related to shrinkage invariably coincide with qualitative change or overall decline in the urban fabric or form of an affected place (see Hollander 2011: 9–10), the latter phenomena can occur in the absence of the former. In other words, decline can be experienced by all types of communities, regardless of whether their population is growing, shrinking, or stable. Decline, it follows, has been actively operating in geographic communities since long before the notion of urban shrinkage came to the attention of academic researchers and planners (Schilling and Mallach 2012). For that reason, the study of urban decline is a relatively mature area of social science inquiry that has been subject to ongoing research for more than a century. By comparison, scholarship on urban shrinkage is still young, having origins that trace mostly to the 1990s (per Beauregard, as quoted by Schilling and Mallach 2012: 24). Nevertheless, literature on both topics – and their many intersections and interrelationships – has exploded in recent decades, with new books and academic articles on "shrinking cities" entering the discourse of academics and practitioners.

The goals of this book are to expand the discussion on patterns of shrinkage and decline in the United States, introduce the difficulty in unpacking the complex processes that have produced these patterns, and examine broader policy frameworks that have been under consideration to address shrinkage in U.S. cities. Chapter 2 describes several indicators of urban shrinkage as well as the data sources from which those indicators can be acquired. Patterns and trends in those indicators are then explored at a variety of geographic levels across the United States. Chapter 3 performs similar analyses on indicators of urban decline. Like all useful attempts to prove the existence of patterns and trends, the main objective of Chapters 2 and 3 is to provide readers with empirical evidence of, and selected analytical tools for studying shrinkage and decline, and the various dimensions of and relationships between them, across the United States. Accordingly, to better situate the empirical findings in urban theory, Chapter 4 synthesizes classic and recent scholarship on factors or processes that lead to shrinkage and decline in central cities. More precisely, the chapter provides a general introduction to theories of intra-urban change. Chapter 5 widens the geographic perspective of that discussion to examine factors that lead to shrinkage and decline beyond central city borders. In that chapter, it becomes clear that "urban" shrinkage and decline are not strictly city-specific issues. Chapter 6 focuses on pro-growth approaches to planning and policymaking. As part of this approach, strategic efforts and investments are geared toward re-growing a shrinking city to its former glory, rather than accepting that such an outcome might be unattainable (Leo and Anderson 2006; Hollander 2011).

In Chapter 7, the book begins to explore opportunities for "rightsizing" (Hummel 2015) by which planners, policymakers, academics, and other stakeholders seek to embrace and accept shrinkage, rather than pursue pro-growth policies to reverse the trend (Haase et al. 2014). Rightsizing therefore involves a major shift in how we perceive and respond to phenomena related to shrinkage and decline. Above all, rightsizing means "planning for less – fewer people, fewer buildings, fewer land uses" (Popper and Popper 2002: 23); which is a clear break from the pro-growth approach in planning and policymaking discussed in Chapter 6. Chapter 8 discusses the trends in metropolitan governance in the United States with a specific focus on how governance is addressing problems related to shrinkage and decline. Specifically, the chapter draws attention to the needs of intergovernmental cooperation, which is a strategy that has received ample attention (and some practical experimentation) in the United States in recent decades. Chapter 9 continues to explore alternatives to pro-growth by introducing the growing planning perspectives of sustainability and resilience. The chapter argues that sustainability and resilience concepts can both reinforce and enhance the rightsizing movement that is introduced in Chapter 7. This chapter offers a discussion on efforts toward creating resilient, livable, and more socially equitable communities. Finally, in addition to summarizing the key issues and ideas from the book, Chapter 10 suggests new directions for research by taking into account recent demographic shifts.

With all the talk of "shrinking cities", it is easy to lose sight of the fact that neither shrinkage nor decline are likely to apply to the whole of any one settlement. Indeed, even cities that have experienced substantial population loss typically feature thriving (even growing) internal neighborhoods where residents have access to ample amenities and enjoy high-quality urban experiences. Accordingly, where possible, this book emphasizes the importance of studying patterns of actually existing shrinkage – that is, the distribution of spaces and places where downward quantitative adjustments in population, the economy, and/or the built environment have in fact occurred. Among other reasons, this spatial perspective has important implications for resource targeting and strategic planning (Schilling and Mallach 2012).

References

Beauregard, Robert A. 2006. *When America Became Suburban*. Minneapolis: University of Minnesota Press.

Beauregard, Robert A. 2009. "Urban Population Loss in Historical Perspective: United States, 1820–2000." *Environment and Planning A* 41 (3):514–528.

Emery, Mary, and Cornelia Flora. 2006. "Spiraling-Up: Mapping Community Transformation with Community Capitals Framework." *Community Development* 37 (1):19–35.

Großmann, Katrin, Marco Bontje, Annegret Haase, and Vlad Mykhnenko. 2013. "Shrinking Cities: Notes for the Further Research Agenda." *Cities* 35:221–225.

Haase, Dagmar, Neele Larondelle, Erik Andersson, Martina Artmann, Sara Borgstrom, Jurgen Breuste, Erik Gomez-Baggethun, Asa Gren, Zoe Hamstead, Rieke Hansen, Nadja Kabisch, Peleg Kremer, Johannes Langemeyer, Emily Lorance Rall, Timon McPhearson, Stephan Pauleit, Salman Qureshi, Nina Schwarz, Annette Voigt, Daniel Wurster, and Thomas Elmqvist. 2014. "A Quantitative Review of Urban Ecosystem Service Assessments: Concepts, Models, and Implementation." *Ambio* 43 (4):413–433.

Hollander, Justin B. 2011. "Can a City Successfully Shrink? Evidence from Survey Data on Neighborhood Quality." *Urban Affairs Review* 47 (1):129–141.

Hollander, Justin B., Karina Pallagst, Terry Schwarz, and Frank J. Popper. 2009. "Planning Shrinking Cities." *Progress in Planning* 72 (4):223–232.

Hospers, Gert-Jan. 2014. "Urban Shrinkage in the EU." In *Shrinking Cities: A Global Perspective*, edited by Harry W. Richardson and Chang Woon Nam. London: Routledge.

Hummel, Daniel. 2015. "Right-Sizing Cities in the United States: Defining Its Strategies." *Journal of Urban Affairs* 37 (4):397–409.

Leo, Christopher, and Kathryn Anderson. 2006. "Being Realistic about Urban Growth." *Journal of Urban Affairs* 28 (2):169–189.

Martinez-Fernandez, Cristina, Ivonne Audirac, Sylvie Fol, and Emmanuèle Cunningham-Sabot. 2012. "Shrinking Cities: Urban Challenges of Globalization." *International Journal of Urban and Regional Research* 36 (2):213–225.

Mumford, Lewis. 1961. *The City in History*. New York: Harcourt, Brace & World.

Myrdal, Gunnar. 1957. *Economic Theory and Underdeveloped Regions*. London: Gerald Duckworth & Co.

Pallagst, Karina, Thorsten Wiechmann, and Cristina Martinez-Fernandez, eds. 2013. *Shrinking Cities: International Perspectives and Policy Implications*. New York: Routledge.

Popper, Deborah E., and Frank J. Popper. 2002. "Small Can Be Beautiful." *Planning* 68 (7):20–23.

Schilling, Joseph, and Jonathan Logan. 2008. "Greening the Rust Belt: A Green Infrastructure Model for Right Sizing America's Shrinking Cities." *Journal of the American Planning Association* 74 (4):451–466.

Schilling, Joseph M., and Alan Mallach. 2012. *Cities in Transition: A Guide for Practicing Planners (Planning Advisory Service No. 568)*. Chicago: American Planning Association.

United Nations. 2014. *World Urbanization Prospects: The 2014 Revision*. New York: United Nations.

2 Patterns and trends

Measuring and mapping urban shrinkage

Chapter 1 defined *shrinkage* as a sustained decrease in the total population of a geographic community. Urban shrinkage is not a novel occurrence but is instead part of a larger trend: shrinkage in one form or another has been operating on the scale of human settlements for centuries (Turchin 2003), and it has been affecting patterns of (de)urbanization in the United States since at least the 1800s (Beauregard 2003, 2009). Also, and perhaps more consequentially, to understand urban shrinkage as a sustained decrease in the number of people living in a particular city, it is necessary to consider population change in conjunction with changes in relevant mediator variables. More precisely, while it is common for variables such as population to experience short-term fluctuations, the presence of long-term shrinkage suggests that there are other factors at work within the urban system (Hollander 2011). The next section explores the various patterns of shrinkage across the United States. The subsections below feature discussions of the data indicators, data sources, and analytical techniques used to explore shrinkage in the United States.

Patterns of shrinkage in the conterminous United States

To study urban change and, by extension, urban shrinkage within a large and diverse study area like the United States, it is typically necessary to define a "universal simple measure" that allows for "comparison in space and time" (Pumain 2006: 6). Arguably the most suitable and readily accessible variable that satisfies this demand is the (changing) population of a geographic unit (Beauregard 2009). Indeed, the size of a place's population has critical implications for what is needed in terms of the place's built environment, production and consumption activities, and public service provision, among other things (Batty 2013).

In a very broad sense, scholars of urban shrinkage use two approaches to identify shrinking places based on population change. First, in what we call the **binary method**, researchers classify a city based on its absolute population change between a given set of time periods. Places where the total population increased or remained static from the first to the last time period analyzed are labeled as growing or stable, while places where the overall population decreased between the same two time periods are categorized as shrinking. The binary method possesses the desirable quality of being non-arbitrary, in that population growth and shrinkage are defined in very precise – and literal – manners. At the same time, the classification scheme does not differentiate between locales experiencing temporary and anomalous population contraction with those undergoing long-term and potentially transformational shrinkage. Therefore, when using the binary method to identify shrinking places, it is important for researchers to give proper consideration to the *prevalence* (the number of times population decreased during the study period), *severity* (the magnitude of population decrease), and *persistence* (the sequential nature of multiperiod population decreases) of any observed population downturns (Beauregard 2009).

The second approach for identifying shrinking places, referred to here as the **threshold method**, relies on a pre-specified critical value of population loss to operationalize shrinkage. Places that experienced population loss greater than or equal to the adopted critical value between a given set of time periods are labeled as shrinking (Schilling and Logan 2008). The threshold method suffers from arbitrariness, insofar as different researchers employ different threshold values, and no optimal threshold is available (or appropriate) to adopt as a bright line test for detecting all urban shrinkage. Nevertheless, the method has considerable utility for academic and policy research. Namely, following the reasoning from the preceding paragraph, the method recognizes that the simple event of population contraction must necessarily be unfolded into more complex considerations of *severity* and *persistence*. More succinctly, the degree and duration of population loss matters. If the magnitude of loss is not *severe* (i.e., at or beyond the adopted threshold), then a place is not classified as "shrinking".

As Hollander (2011) and others (Glaeser and Gyourko 2006; Weaver and Knight 2014) have pointed out, cities that are built for a certain level of population tend to experience metamorphic (frequently negative) physical and economic changes following severe and persistent population loss. The same can rarely be said for cities that endure comparatively marginal or temporary reductions in population. These issues are addressed further in Chapter 3, which explores patterns of *urban decline*. For now, it is sufficient to conclude this subsection by noting two of the most commonly cited thresholds employed by scholars in the shrinking cities literature. First, Schilling and Logan (2008) define shrinkage as a 25 percent or greater loss in a place's total population over the course of four decades. Second, Hollander (2011) adopts the slightly more conservative, but quite similar, threshold of 30 percent over four decades. While, again, these thresholds are somewhat arbitrary, they are values that have been put forward by urban planning and policy experts to define the parameters of population shrinkage based on their collective knowledge and experience. As such, these thresholds will be referenced throughout the remainder of this chapter.

Population data for the United States

The premier source of population (and many other) data for the United States is the U.S. Census Bureau (see Box 2.1 on accessing Census Bureau data). While the Census Bureau creates and distributes a wide variety of products and tools that describe

Box 2.1 Accessing population and other data for the United States

Census and other urban data, as well as geographic boundary files for use in geographic information systems (GIS), are obtainable from a number of online sources. Free sources include the Census Bureau's American FactFinder, National Historic Geographic Information System, and the Longitudinal Tract Data Base. Commercial sources requiring a paid subscription or license include Social Explorer and Esri's Business Analyst and Community Analyst.

The Census Bureau conducts nearly one hundred surveys and censuses every year. By law, no one is permitted to reveal information from these censuses and surveys that could identify any person, household, or business. Data from the following surveys and censuses are available in American FactFinder:

American FactFinder (AFF) is the U.S. Census Bureau's official online data distribution portal for all Census Bureau surveys dating back to 2000. The Census conducts and reports data for more than 100 surveys and censuses, including the Decennial Census, American Community Survey, Economic Census, Census of Governments, and American Housing Survey. Given the depth of surveys and censuses, available data includes demographics and socio-economic; housing; industrial and business; and employment. Although not available directly in AFF, the Census also offers geographic boundary files for GIS. Since the census is both collector and reporter, one of the key benefits of the AFF is that metadata and resources are provided. Users can select data by census geography, program, year, and variable, allowing for custom downloads of geographically specific data to the block level. For those seeking current data for a specific geography, such as a zip code, the Community Facts search capability provides quick and easy access to frequently requested population, income, and housing data. AFF also provides limited online mapping, for example the Census Flows Mapper which allows access to county-to-county migration data. However, better online mapping capability is available with other, albeit fee-based, resources including Social Explorer or Esri's Business Analyst or Community Analyst.

The Minnesota Population Center at the University of Minnesota hosts and maintains the **National Historical Geographic Information System** (NHGIS). This excellent free online source extends the data available in American FactFinder to include data and boundary files for decennial census data for all geographies from the original 1790 census to the present. The Data Finder allows users to filter their data search by geographic level, years, topics, and database. One useful feature is that once the data have been selected, the user can also select and download the correct geographic boundary file in the same request. Recently, the NHGIS estimated and made available block-based data for ten geographies from the 2000 census standardized to 2010 boundaries, offering researchers the ability to analyze from 2000 to 2010 change more accurately.

For researchers seeking to analyze decadal change going back further than 2000, the **Longitudinal Tract Data Base** (LTDB) at Brown University provides the ability to do so back to 1970. LTDB was developed to help researchers tackle one key issue with respect to analyzing historic data – the modifiable areal unit problem (MAUP). Simply put, the boundaries of census geographies change over time, limiting the ability of researchers to track changes within the same geographic area over one or more census periods. The LTDB estimates common data variables for prior census tract-level data within the 2010 census tracts, which users can download for the 1970–2000 Decennial Censuses. Users with their own data aggregated to census tracts can use the LTDB to estimate this data within the 2010 tract boundaries.

Social Explorer offers a wealth of downloadable data and an excellent mapping interface, but the fully functional site requires a paid subscription. The available data includes all U.S. decennial censuses from 1790 to 2010 and all American Community Survey data from 2005 to present. Social Explorer has two significant advantages. The first is a user-friendly interface that allows new users to quickly and easily select and download data. The data selection interface is nearly identical to the popular prior version of the AFF, utilizing dropdown menus to select and download data in common database formats, including Excel. The second advantage is the mapping tool, which allows users to easily map variables to visualize data spatially. The mapping tool also allows the mapping of multiple variables as well as a side-by-side mapping capability allowing a user to map variables from two different censuses side by side in the same geography.

Esri's ***Business Analyst*** and ***Community Analyst*** are two subscription-based tools for geographically focused, data-driven, strategic decision-making. Each has ostensibly the same data but is intended for two different types of users. Business Analyst is geared toward real estate, economic development, retail, and marketing. In Business Analyst, users can download existing or create custom summary reports and maps of business counts, employees, total sales, consumer spending, and the market potential of a given area for certain goods and services. Community Analyst is aimed at policymakers and planners and has much of the same data and functionality. The value of these databases lies in their ability to create ready-made maps and reports based on user-defined geographies, eliminating the need to download, analyze, and report the data oneself, as one would from the census or NHGIS. An additional benefit is that each offers five-year and ten-year projections for common variables such as population, households, income, and housing.

characteristics of the people, housing units, economy, and government structure of the United States (MacDonald and Peters 2011), the data and visualizations presented here rely exclusively on two of these items: (1) the decennial census and (2) the Topologically Integrated Geographic Encoding and Referencing (TIGER) products.

By Constitutional mandate, the Census Bureau is required to perform a full population count for the United States at least once every ten years to facilitate the processes of political reapportionment and redistricting (Bullock 2010). Hence, since 1790, the Bureau has conducted a population census once every decade, such that these **decennial censuses** correspond to years ending in zero. Among other variables, each decennial census (also called the "full count") includes data on the number of people, families, households, housing units, [owner- and renter-] occupied housing units, and vacant housing units for a variety of geographic levels of analysis. Further, the majority of these variables can be broken out by age, age group, gender, race, and ethnicity (though, with respect to the latter, refer to Yanow's 2003 book on how the Census Bureau's race and ethnicity constructions severely limit one's ability to report one's own, self-authored identity to governmental data collectors). For current purposes, the decade-to-decade changes in a place's total population that are captured in the decennial censuses offer, as argued above, a "universal simple measure" for studying patterns and trends in urban shrinkage (Pumain 2006): *universal* in the sense that the decennial census is conducted and reported on in a uniform fashion for the whole United States; and *simple* in that population growth or contraction is an obvious indicator of place-based change (Beauregard 2009).

That said, if the goal is to explore patterns and trends in decadal population loss in the United States, then it is necessary to specify *where* in the United States population loss can be studied. Answering this "where" question calls for a basic understanding of the **geographic framework** of the U.S. Census Bureau. Table 2.1

Table 2.1 Basics of the U.S. Census geographic framework

	Statistical Areas	*Political Areas*
Coarse resolution	Nation	Nation
Fine resolution	Census region	State and territory
	Census division	Congressional district
	Core-based statistical area	County
	Census tract	Vote tabulation district
	Census block group	Place
	Census block	American Indian area
Nesting	Units listed above are perfectly nested – i.e., higher levels completely contain lower levels	Units listed above are not perfectly nested
Relationships to the other area types	Core-based statistical areas fully contain counties, which fully contain census tracts and all sub-tract units	Counties fully contain census tracts and all sub-tract statistical areas; however, places can cross the borders of core-based statistical areas, counties, census tracts, and all sub-tract statistical areas

presents a non-exhaustive summary of this framework, where geographic units of analysis are classified into *political areas* and *statistical areas*. **Political areas** are units of analysis that correspond to political jurisdictions. They include areas such as the nation as a whole, congressional districts, states and territories, counties, and places. Of these political areas, *places* are the closest correspondent to municipalities (e.g., cities and towns). Places include both *incorporated places* as well as *census designated places*, which are unincorporated but identifiable by name.[1] Importantly, because there is "no . . . centralized source of data on [municipal] boundaries" in the United States, scholars often rely on the census *place* geography to study phenomena (e.g., population shrinkage) at the city level (Kodrzycki and Muñoz 2015: 114).

Statistical areas are spaces delineated by the Census Bureau, typically in conjunction with state and local officials, for ongoing data collection purposes. Crucially, although persons are surveyed at their residential addresses, and census data are therefore collected at the household level, the Census Bureau only reports data for aggregated spatial units to protect respondent confidentiality. The smallest statistical area, a census block, facilitates this objective. The census block is used as the basic building block for constructing all other statistical areas. As the following section discusses in more detail, the Census Bureau's rolling American Community Survey (ACS) collects data on various socioeconomic characteristics of the population. Because these data are based on a sample of the population, and as a further guard on the confidentiality of sensitive socio-economic information, they are reported for units of analysis larger than census blocks. Most (but not all) such data are available for small clusters of census blocks, which are fittingly called *census block groups*; and all of the ACS data are generally available for census tracts, which are slightly larger, and strictly contain census block groups. Despite being imperfect proxies for neighborhoods (Kwan 2012), census tracts are generally used to study phenomena at a "neighborhood" level (Hollander 2011). When research questions are more macroscopic in nature, census core-based statistical areas (CBSAs), which encompass metropolitan and micropolitan areas, are typically used as proxies for *urban regions* that include both a city and its surrounding communities. Finally, the Census Bureau divides the United States into four distinct, large geographic regions – West, Midwest, South, and Northeast – that are in turn divided into nine geographic divisions (Fig. 2.1). These multistate areas allow researchers to track broad spatial trends in population and economic phenomena over time, as demonstrated in in the following section.

For all spatial units included in the census geographic framework, each individual geographic area is assigned a unique code based on the Federal Information Processing Series (FIPS). These FIPS codes are what link tabular census data to locations in space. In order to spatialize census data within a geographic information system (GIS), the Census Bureau provides TIGER cartographic boundary files for each of its various geographic levels of analysis. The TIGER files are available as shapefiles (a spatial data format native to Esri software), geodatabases (a spatial database format native to Esri software), and KML files (a spatial layer format native to Google Earth). Every spatial entity included in the TIGER files is linked to a unique FIPS code, meaning that TIGER spatial data can be joined with tabular census data (e.g., the full count dataset) in a GIS.

One of the major challenges of performing multitemporal spatial analysis with census data is that cartographic boundaries frequently change. Boundaries of statistical areas such as census tracts tend to change with each decennial census, while political boundaries, such as place borders, can change through annexation or various other processes. Accordingly, studying patterns of population change over time in the United States requires one to normalize these changing geographies of data collection. A dataset produced by Logan, Xu, and Stults (2014), which is freely accessible through Brown University (see Box 2.1), provides interpolated data from past decennial censuses for current (2010) census tract boundaries. More explicitly, the Brown University dataset provides users with high quality approximations of key variables from the 1970, 1980, 1990, and 2000 decennial censuses – as well as the corresponding (known) variable values from the 2010 census – all within current census tract boundaries. The data therefore allow for consistent "comparison in space and time" (Pumain 2006: 6). However, because the entire United States was not yet "tracted" in 1970 – that is, by 1970 the census tract system had not yet been implemented for the entire country, only for mostly urbanized areas – analyses that wish to study population change going back to 1970 are limited to a subset of mostly urban locations throughout the United States This subset of locations is depicted in Figure 2.1 for the conterminous United States. Figure 2.1 also pictures the four U.S. Census regions and their nine constituent

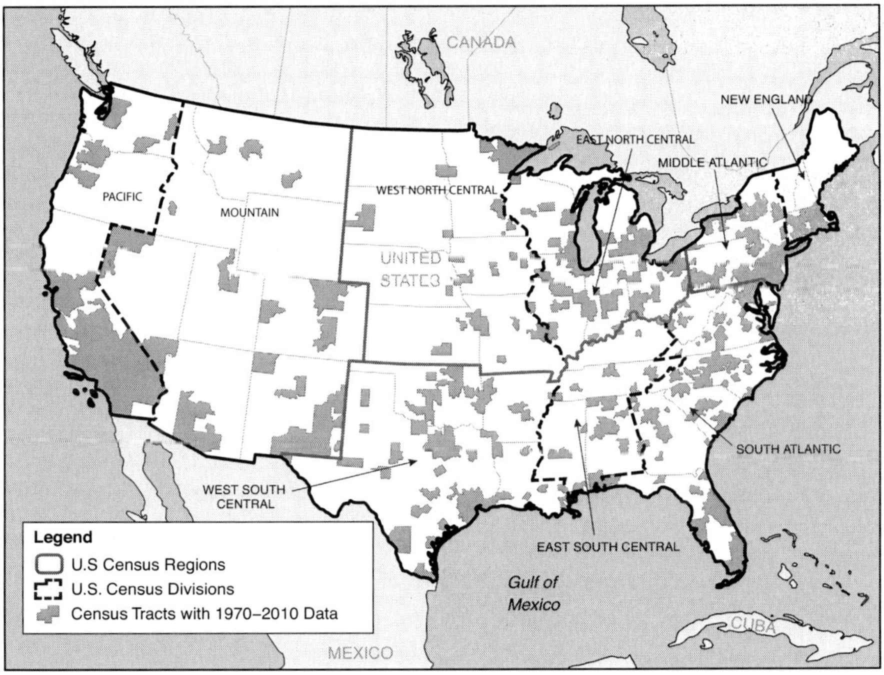

Figure 2.1 Census tracts included in the Brown University dataset, for the conterminous United States, by U.S. Census region and division

divisions, which are used below to explore macro trends in patterns of shrinkage: (1) the West region is made up of the (i) Pacific and (ii) Mountain divisions; (2) the South region consists of the (iii) West South Central, (iv) East South Central, and (v) South Atlantic divisions; (3) the Midwest region includes the (vi) West North Central and (vii) East North Central divisions; and (4) the Northeast region consists of the (viii) Middle Atlantic and (ix) New England divisions.

The next subsection draws on the Brown University dataset to analyze population shrinkage in the United States since 1970 for the collection of census tracts pictured in Figure 2.1. Alaska and Hawaii are excluded from the analysis on the grounds that their relatively recent statehood (1959) precluded them from participating in the formative years of major federal urban policy initiatives, such as the Federal Highway Act (Kinevan 1950). Because these federal policies played major roles in shaping patterns of post-World War II urban development, the initial ineligibility of non-state territories to participate in these programs is sufficient justification for limiting the analysis to the conterminous states.

Analysis of population change, 1970–2010

This subsection presents results from a series of related analytical operations that are collectively intended to describe geographic patterns and trends in urban population shrinkage in the United States. While each analytical operation requires its own caveats, which are made explicit at appropriate points below, three global study design choices apply to all of the analyses. First, the full temporal (1970–2010) and spatial (Fig. 2.1) extents of the empirical exercises are determined by the data currently available in the Brown University LTDB (see Box 2.1). Second, and related, the census tract is the foundational unit of analysis employed throughout the subsection. Analyzing population change at the tract level contributes to an understanding of *actually existing urban shrinkage*, rather than wholly classifying cities as "shrinking" without any consideration of intra-city patterns of population change. As such, tract-level studies regularly demonstrate that even so-called "shrinking cities" contain growing and thriving neighborhoods (Hollander 2011). Third, at least for the purpose of classifying census tracts, the analyses implement the *threshold method* of identifying shrinkage as described earlier in this chapter. Following existing studies (Schilling and Logan 2008), the adopted threshold is 25 percent population loss over forty years (i.e., from 1970–2010). Although this threshold – and all thresholds for that matter – are arbitrary, the 25 percent critical value features in one of the most influential recent articles in the shrinking cities literature, which was authored by planning experts with extensive experience in and knowledge of the topic (Schilling and Logan 2008).

With these global research design choices in mind, the remainder of this subsection attends to the following five questions:

1. Where are the census tracts that meet the adopted threshold for shrinkage located?
2. Where are census tracts that appear to be on target to meet the adopted forty-year threshold for shrinkage over the next ten, twenty, and thirty years located?

3 Where are the census *places*[2] that meet the adopted threshold for shrinkage located?
4 Are there growing (shrinking) tracts within shrinking (growing) places?
5 Are shrinking census tracts found disproportionately in shrinking places? Does this pattern appear to be changing with time?

Before engaging directly with these research questions, the adopted shrinkage threshold of 25 percent population contraction over forty years is used to derive a threshold **exponential rate of population change**. A common assumption in population geography is that the rate of population change in a given spatial unit is "an instantaneous one that may be applied continuously" (Plane and Rogerson 1994: 59). Such an outcome can be described with an exponential population (de) growth function:

$$P_t = P_0 e^{rt}, \qquad [1]$$

where P_t is the population of the geographic unit after t time periods, P_0 is the initial population, e is Euler's constant (approximately equal to 2.718), and r is the continuous rate of change. To solve for the annual rate of change (r) that produces a 25 percent population loss over forty years, Equation 1 can be algebraically rearranged to yield:

$$r = \left[ln\left(\frac{P_t}{P_0}\right) t^{-1} \right] * 100. \qquad [2]$$

Observe now that a 25 percent population loss over four decades implies that a geographic unit's population at the end of forty years is equal to 75 percent of its initial population. As such, one can solve for the appropriate annual (threshold) *rate* of exponential population change as:

$$r = \left[\frac{ln(0.75)}{40} \right] * 100 \cong -0.719. \qquad [3]$$

In other words, census tracts (or places) whose populations contract at a rate of at least −0.719 percent per year will shed a minimum of 25 percent of their population over a forty-year period. This rate becomes especially important in addressing question #2 from above, which concerns the locations of census tracts that appear to be *on target* to meet the adopted threshold to be considered "shrinking" in the coming decades.

Beginning with question #1, Figure 2.2 maps the distribution of census tracts from the Brown University LTDB whose annual rates of exponential population loss from 1970 through 2010 exceed the adopted threshold of −0.719 percent per year. In other words, the highlighted tracts are those that shed 25 percent or more of their population over the course of the four most recent decennial censuses. Using absolute population loss as a weight, Figure 2.2 also depicts the population shrinkage centroid, or weighted geographic mean center of the shrinking tracts.

Following Plane and Rogerson (1994: 31), the **population shrinkage centroid** is computed as:

$$\bar{x} = \frac{\sum_{i=1}^{n} L_i x_i}{\sum_{i=1}^{n} L_i}, \ \bar{y} = \frac{\sum_{i=1}^{n} L_i y_i}{\sum_{i=1}^{n} L_i}, \qquad [4]$$

where $\bar{x}$ and $\bar{y}$ are the coordinates of the shrinkage centroid, L_i is the absolute value of population that was lost in tract i between 1970 and 2010, and x_i and y_i are the coordinates of tract i's centermost point. It should be noted that none of the tracts shown in Figure 2.2 experienced population growth during this timeframe.

In line with Hollander's (2011) findings that population shrinkage is not strictly a "Rust Belt" phenomenon – where researchers conventionally place the Rust Belt in the Midwest and Northeast Census Bureau regions – Figure 2.2 shows that shrinking tracts are found throughout the conterminous United States. At the same time, the "center of gravity" (Plane and Rogerson 1994: 48) of population shrinkage with respect to the 25 percent threshold, as given by the population shrinkage centroid located in Indiana in Figure 2.2, strongly suggests that, on average, long-term population loss has been more *prevalent* and *severe* in the Rust Belt relative to the rest of the country (Beauregard 2009). One can reach this conclusion by

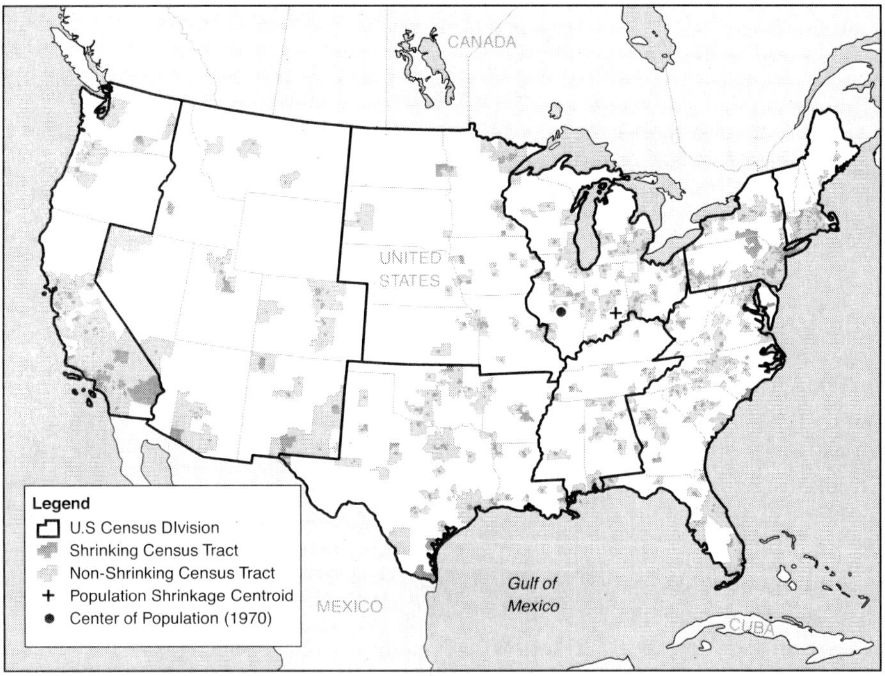

Figure 2.2 Distribution and loss-weighted centroid of shrinking census tracts in the United States (n = 7,386 shrinking tracts)

comparing the location of the *population shrinkage centroid* to the mean center of population in 1970 (the latter is symbolized as a dark circle located in Illinois in Fig. 2.2). If census tracts experienced population change evenly across the study area, then the population shrinkage centroid would lie on top of the mean center of the 1970 population. In Figure 2.2, though, the former is located approximately 190 miles (as the crow flies) to the east of the latter, which suggests that the magnitude of population loss from 1970–2010 was greater in tracts in the eastern part of the country. Hence, in response to question #1, one can say that the distribution of shrinking census tracts in the United States is geographically dispersed; but, aligning with common perceptions, population shrinkage has been more pronounced in the Midwest and Northeast compared to the South and West.

Regarding question #2, the threshold [annual] rate of exponential population change (-0.719 percent per year) can be used to identify census tracts that are *on pace to be shrinking* in the coming decades. In other words, the rate (r) from Equation 2 can be computed for each tract for four possible starting years: (1) 1970, which has already been done to create the distribution pictured in Figure 2.2, and for which t was set equal to 40; (2) 1980, for which t is set equal to 30 (i.e., there are 30 years between 1980 and 2010); (3) 1990, for which t is set equal to 20; and (4) 2000, for which t is set equal to 10. The resulting rates can then be compared to the threshold to classify tracts as being "on pace" to experience a 25 percent or greater loss in population over forty years.

To understand how this classification process works, consider the four hypothetical census tracts described in Table 2.2. Data collected for Tract A begin in 1970 with an initial population of 1,000 persons and end in 2010 with a current population of 700 persons – a 30 percent contraction. The annual rate of exponential population change between 1970 and 2010 is therefore -0.892 percent, which exceeds the threshold and qualifies this tract as *already shrinking* by 2010. Tract B starts with a population of 1,000 and experiences [mostly] continuous population loss but only to the tune of 13.5 percent over the four-decade time period. As such, the annual exponential rate of population change in Tract B from 1970 to 2010 is −0.363 percent, which does not meet the threshold for labeling the tract as *already shrinking*. However, upon closer inspection, one can see that the annual rate of exponential population change in Tract B from 1980 and 2010 was −0.740 percent – greater than the threshold rate of −0.719 percent. For this reason, Tract B is *on pace* to be classified as shrinking by 2020. Similar considerations show that, although Tract C and Tract D *gained* population between 1970 and 2010, they are currently *on pace* to be classified as shrinking by 2030 and 2040, respectively. Indeed, by 2010 Tract C lost 22.7 percent of its 1990 population $\left(\frac{1160-1500}{1500}\right)$; and, despite growing by 120 percent between 1970 and 2010 $\left(\frac{2200-1000}{1000}\right)$, Tract D shed roughly 10 percent of its population between 2000 and 2010. Accordingly, notwithstanding their historical population *growth*, researchers might wish to flag these tracts for exhibiting more recent tendencies toward *shrinkage*.

Figure 2.3 maps the distribution of all census tracts from the Brown University LTDB that are *already shrinking* or *on pace to be shrinking* over the next ten to thirty years, where the latter classifications follow from the reasoning and

Table 2.2 Analyzing the relationship between population shrinkage and economic shrinkage

Year (Y)	*Tract A Pop.*	$r_{A,Y}$ [a]	*Tract B Pop.*	$r_{B,Y}$ [a]	*Tract C Pop.*	$r_{C,Y}$ [a]	*Tract D Pop.*	$r_{D,Y}$ [a]
1970	1000	–0.892%*	1000	–0.363%	1000	0.371%	1000	1.971%
1980	995	–1.172%	1080	–0.740%*	1200	–0.113%	2000	0.318%
1990	860	–1.029%	999	–0.720%	1500	–1.285%*	2445	–0.528%
2000	770	–0.953%	933	–0.757%	1275	–0.945%	2440	–1.035%*
2010	700	n/a	865	n/a	1160	n/a	2200	n/a

Summary:

- Tract A is *already shrinking* (crossed threshold in 1970)
- Tract B is *on pace* to be shrinking by 2020 (crossed threshold in 1980)
- Tract C is *on pace* to be shrinking by 2030 (crossed threshold in 1990)
- Tract D is *on pace* to be shrinking by 2040 (crossed threshold in 2000)

[a] $r_{i,Y} = [ln\left(\frac{P_{i,2010}}{P_{i,Y}}\right) / (2010 - Y)]$, i:{A, B, C, D}, Y:{1970, 1980, 1990, 2000}; * indicates r crossed the adopted annual population change threshold in census year Y

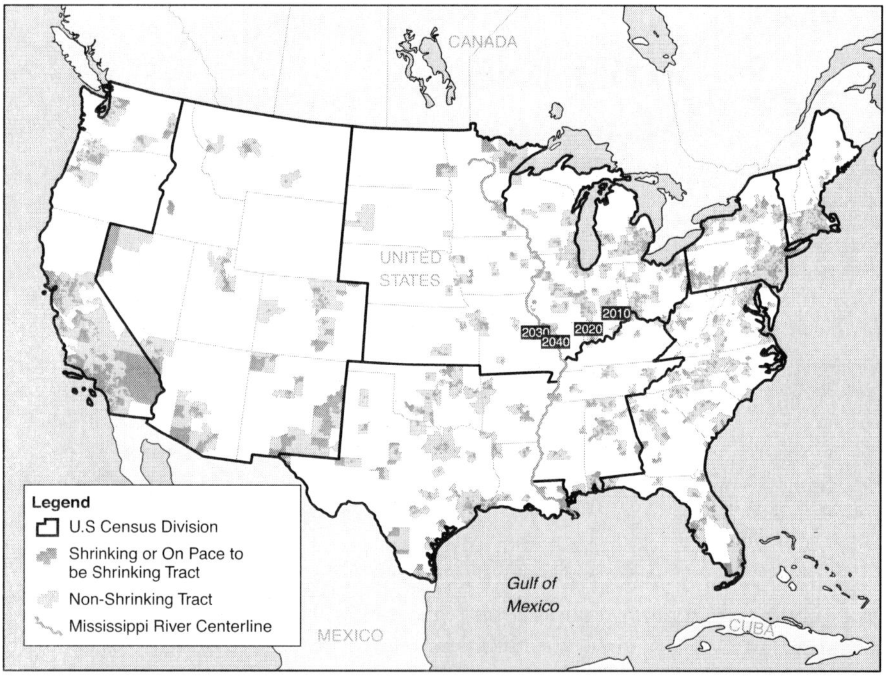

Figure 2.3 The geographies of all "already shrinking" and "on pace to be shrinking" census tracts

examples laid out above. Given the fine grain of a census tract and the substantial extent of the conterminous United States, Figure 2.3 does not make an attempt to visually distinguish between already shrinking and soon-to-be shrinking tracts. Instead, the population shrinkage centroid is used to highlight how the center of gravity of urban shrinkage is likely to shift in the coming decades based on observable rates of population change. Whereas the current population shrinkage centroid (labeled "2010" in Figure 2.3) is well within the geographies of the Rust Belt (Beauregard 2009), this location is moving rapidly to the south and west. In fact, the shrinkage centroid for tracts that have been losing population at a rate of −0.719 percent or greater since 1990 (and thus are on pace to be classified as shrinking by 2030) lies fully to the west of the Mississippi River. Inasmuch as the Mississippi River marks a commonly used heuristic for dividing the conterminous United States into "east" and "west" sections, the evidence presented in Figure 2.3 offers reason to believe that population shrinkage is on track to become an increasingly "western" phenomenon in the United States.

That the population shrinkage centroid is moving to the south and west supports the thesis that the American "Sun Belt" (see Hollander 2011: 48 for a common cartographic depiction of the Sun Belt) is experiencing emerging patterns of severe and persistent population contraction. To explore these changing patterns in more detail, it is possible to measure the degree to which population shrinkage is, and is on pace to be, *concentrated* in or evenly distributed between the nine different U.S. Census Divisions shown in Figure 2.1. A location quotient is a parsimonious indicator of how concentrated a given phenomenon is or is not in the geographic units that make up a larger study area. With respect to shrinking census tracts, a simple **location quotient** can be computed for each large area Census Division (Fig. 2.1) as:

$$Q_i = \frac{s_i / n_i}{S / N}, \qquad [5]$$

where Q_i is the location quotient for Census Division i, s_i is the number of tracts classified as shrinking in Census Division i, n_i is the total number of tracts in Census Division i, S is the number of tracts classified as shrinking in the full dataset, and N is the total number of all tracts in the dataset. Insofar as Q_i is a ratio of ratios, it is a continuous, unitless measure that takes on non-negative values. A location quotient equal to 1 means that the fraction of tracts classified as shrinking in Census Division i is exactly equal to the fraction of tracts classified as shrinking in the study area as a whole. Values of Q_i equal to 1 therefore imply that Census Divisions i contain a proportionate share of the nation's shrinking tracts. Along those lines, when a location quotient takes on a value other than 1, the distribution of population shrinkage is not evenly spread across Census Divisions. A location quotient less than 1 indicates that a Census Division has a lower share of shrinking tracts relative to the study area; and a value greater than 1 implies that shrinking tracts are relatively *concentrated* in the given Census Division.

Figure 2.4 plots the location quotients, by Census Division (Fig. 2.1), associated with currently shrinking tracts as well as tracts that are on pace to be classified as shrinking in each of the next three decades (Fig. 2.3). The pattern of results reaffirms that the geographies of population loss are in flux in the United States, and that population shrinkage is beginning to be distributed more between the Rust Belt and Sun Belt (Hollander 2011). Whereas already shrinking tracts – those that lost 25 percent or more of their population over the last four decades – are expectedly concentrated in the Northeast and Midwest, i.e., "Rust Belt" regions (in addition to the Deep South and Appalachian states that make up the "East South Central" Census Division); tracts that are on pace to be shrinking by 2020, 2030, and 2040 are far more dispersed across the country. For instance, if current rates of population change persist into the future, then Mountain states will experience a sharp increase in their historically low shares of shrinking tracts, while the shares

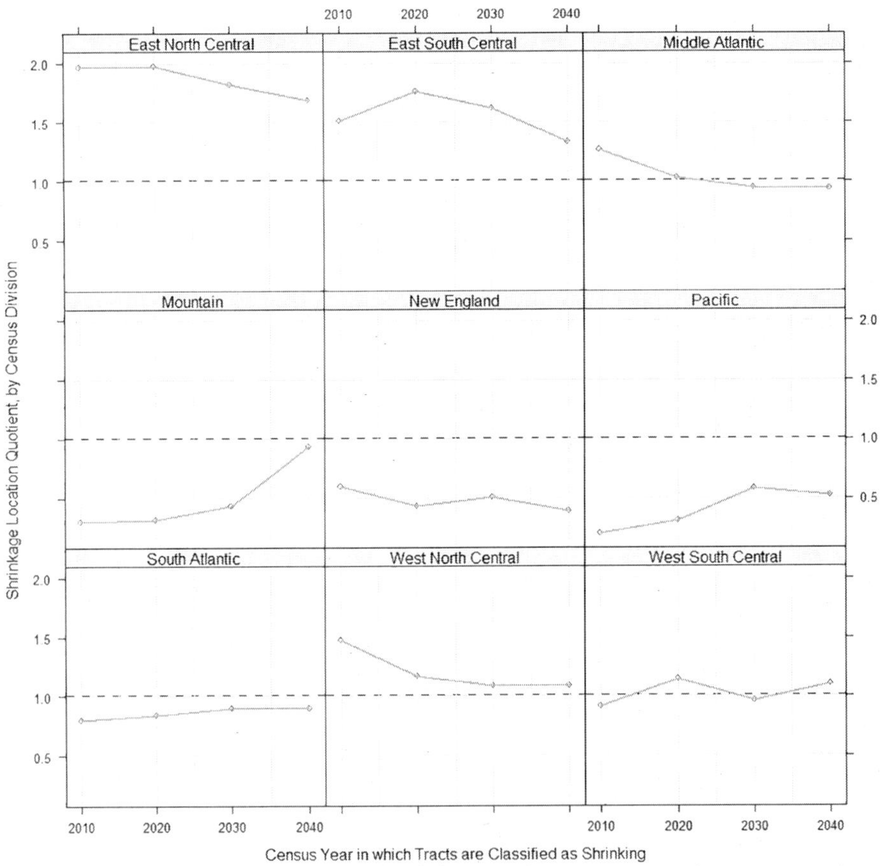

Figure 2.4 The changing concentration of "already shrinking" and "on pace to be shrinking" tracts in U.S. Census Divisions as measured by location quotients

of shrinking tracts in the Middle Atlantic and West North Central portions of the Rust Belt will, for the first time in recent history, become relatively representative of the nation as a whole.

At this point it is clear that *actually existing urban shrinkage* in the United States applies neither to whole cities nor exclusively to locations in the Midwest and Northeast. Nevertheless, images of a shrinking Rust Belt likely endure because those cities or municipalities where aggregate population loss has been most *severe* and *persistent* are more *prevalent* in this region compared to other parts of the country (Schilling and Logan 2008; Beauregard 2009; Schilling and Mallach 2012). On that note, and in the context of the third research question posed above (Where are the census *places* that meet the adopted threshold for shrinkage located?), Figure 2.5(a) maps the distribution of *census places* with a minimum population of 50,000 in the 2010 decennial census. Highlighted in the map is the subset of these places that lost population at an annual rate of −0.719 percent (the adopted threshold for *shrinkage* from above) or greater over the last four decades.

Because the boundaries of census places can change quite frequently, and since the Brown University LTDB provides historical (geographically normalized) data exclusively for current *census tracts*, analyzing population change for *census places* is somewhat challenging. For this exercise, the decision was made to aggregate tract-level data from the Brown University LTDB to current (2010) census place boundaries. However, because tracts are not all fully contained by place boundaries, this aggregation process is necessarily imprecise. For instance, to create Figure 2.5, tracts were aggregated to places on the basis of the formers' centermost points. That is, each tract was assigned to the place (minimum population of 50,000 in 2010) that contains the tract's geographic centroid. A half-mile buffer from place boundary to tract centroid was applied to this join process to maximize the likelihood that the full extent of a place would be captured in the analysis – as many places are intersected by irregularly shaped tracts that lie mostly within the place boundaries, but their centroids fall outside those boundaries.

Figure 2.5 contributes to the existing shrinking *cities* literature in two important ways. First, it expands the scope of the analysis to include all census places that had a minimum population of 50,000 persons in 2010. This expansion is important insofar as 50,000 persons is used as a marker of eligibility to receive funding from many federal urban initiatives. In other words, places with at least 50,000 persons are generally considered *cities* by the U.S. government. Prior studies have tended to omit these smaller cities from consideration, focusing instead on places with 100,000 or more persons (Hollander 2011), or those that crack "top-50" population lists in a given decade (Beauregard 2009). Second, by applying the same analytical techniques used to address research questions #1–2 above, panel (b) in Figure 2.5 shows the places that are on track to experience a 25-percent drop in population between 2000 and 2040. In other words, Figure 2.5(b) maps the distribution of places that are on pace to be classified as shrinking in the next thirty years.

Table 2.3 enumerates the specific places (cities) highlighted in the two panels of Figure 2.5. Observe that with the exception of New Orleans and, more recently, a handful of other Gulf Coast communities whose populations were presumably

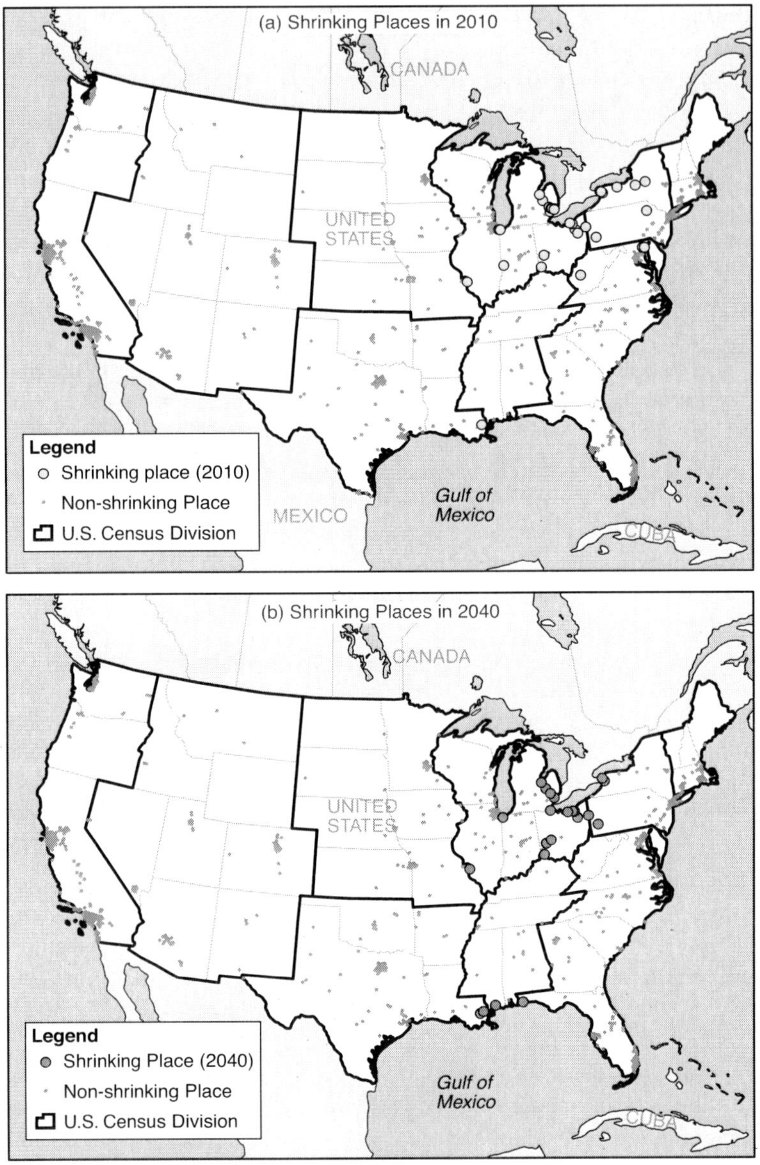

Figure 2.5 Distribution of "already shrinking" places and places "on pace to be shrinking" by 2040

negatively affected by Hurricane Katrina, *shrinking cities* do indeed seem to be a "Rust Belt" phenomenon. Moreover, unlike the shifting patterns of *actually existing urban shrinkage* that are described in Figures 2.3 and 2.4, the spatial concentration of shrinking cities in the Rust Belt is not expected to change significantly in the coming decades (Fig. 2.5b). While far fewer places are on target to experience

Table 2.3 List of "already shrinking" and "on pace to be shrinking" places (cities) in the conterminous United States

Already Shrinking Places[a,b]	*Places On Pace to be Shrinking in 2040*
Akron, Ohio	Akron, Ohio
Baltimore, Maryland	Buffalo, New York
Buffalo, New York	Cincinnati, Ohio
Canton, Ohio	Cleveland, Ohio
Charleston, West Virginia	Dayton, Ohio
Cincinnati, Ohio	Detroit, Michigan
Cleveland, Ohio	Flint, Michigan
Dayton, Ohio	Gary, Indiana
Dearborn Heights, Michigan	**Gulfport, Mississippi***
Detroit, Michigan	Lakewood, Ohio
Dundalk, Maryland	**Lorain, Ohio**
Flint, Michigan	**Metairie, Louisiana***
Gary, Indiana	New Orleans, Louisiana*
Hammond, Indiana	Niagara Falls, New York
Lakewood, Ohio	**Pensacola, Florida***
New Orleans, Louisiana*	Pittsburgh, Pennsylvania
Niagara Falls, New York	Pontiac, Michigan
Pittsburgh, Pennsylvania	Saginaw, Michigan
Pontiac, Michigan	**Springfield, Ohio**
Rochester, New York	St. Louis, Missouri
Royal Oak, Michigan	**Toledo, Ohio**
Saginaw, Michigan	Youngstown, Ohio
Scranton, Pennsylvania	n = 22
St. Claire Shores, Michigan	
St. Louis, Missouri	
Syracuse, New York	
Terre Haute, Indiana	
Tonawanda (city), New York	
Utica, New York	
Warren, Michigan	
Youngstown, Ohio	
n = 31	

[a] Minimum population of 50,000 persons in 2010; [b] Place-level populations were computed using census tracts whose centermost points lie fully in 2010 place boundaries; *italicized text* in the first column indicates places that have not been losing population at or above the adopted threshold rate since 2000, and thus do not appear in the second column; **bold text** in the second column indicates places that do not appear in the first column; * indicates Sun Belt state on the Gulf Coast

25 percent or greater population loss between 2000 and 2040 relative to the number of cities that shrank between 1970 and 2010 (Table 2.3), the distribution of these places remains highly concentrated in Rust Belt states, especially Michigan and Ohio.

To round out the population analysis (questions #4–5 from above), Table 2.4 summarizes the distribution of shrinking tracts and tracts that are projected to be

Table 2.4 Distribution of "shrinking" and "on pace to be shrinking" tracts by their location in shrinking or non-shrinking places

	Shrinking Place (Current)	*Non-Shrinking Place*
Shrinking Tract (Current)	1,582	3,867
Non-Shrinking Tract	664	24,796
» Shrinking places account for 7.27% of all tracts, but 29.03% of shrinking tracts		
Shrinking Tract (2020)	1,330	3,098
Non-Shrinking Tract	916	25,565
» Shrinking places account for 7.27% of all tracts, but 30.04% of shrinking tracts		
Shrinking Tract (2030)	1,274	3,380
Non-Shrinking Tract	972	25,283
» Shrinking places account for 7.27% of all tracts, but 27.37% of shrinking tracts		
Shrinking Tract (2040)	1,323	5,186
Non-Shrinking Tract	923	23,477
» Shrinking places account for 7.27% of all tracts, but 20.33% of shrinking tracts		

Notes: (1) Chi-squared tests on all four contingency tables are significant with $p << 0.001$, suggesting that shrinking tracts are disproportionately concentrated in shrinking places; (2) of the 52,380 conterminous census tracts included in the Brown University LTDB, n = 30,909 have their centroids in places with a minimum (2010) population of 50,000 persons

shrinking in the next ten to thirty years, according to their location in the (non-) shrinking places pictured in Figure 2.5(a). While it is quite clear that shrinking places contain a disproportionate share of these shrinking and soon-to-be shrinking tracts in the overall study area, there are signs that a reversal in this trend could be on the [distant] horizon. For instance, whereas 70.44 percent of tracts in shrinking places shrank between 1970 and 2010 (1,582 of 2,246), if current rates of population change continue, then this rate will fall to 58.90 percent (1,323 of 2,246) by 2040 – an 11.5 percentage point drop. Meanwhile, the corresponding change in non-shrinking places is a 4.6 percentage point *gain* in shrinking tracts. Thus, one can conclude that shrinking places are not wholly comprised of shrinking tracts (question #4). Furthermore, the fraction of tracts classified as shrinking in such places is on track to decrease markedly in the coming decades (question #5).

Per capita and aggregate income

Another important subject to understand shrinkage is the wealth of a geographic unit. Wealth in a community exists in multiple dimensions, some of which (e.g., financial capital) are tangible, and others, such as social capital, are intangible (Flora, Flora, and Gasteyer 2015). It is beyond the scope of this subsection to examine these various forms of community wealth. Instead, here it is claimed that *per capita income* and *aggregate income* are [imperfect] indicators of the financial wealth present within a given geographic unit at a specific point in time.[3]

Data on the income level and other socio-economic characteristics of people in the United States are available through two main Census Bureau surveys. Unlike the decennial censuses, which aim to be "full counts" of the population, census survey data are compiled from samples. The first important source of sample-based data discussed here is the Census Summary File 3 (SF3), which contains data on respondents' income, education, migration, household costs, and living arrangements. Up to and including the year 2000, the SF3 was distributed to approximately one in six U.S. households as a long form attachment to the standard (i.e., short form) decennial census. Since 2005, however, the SF3 has been replaced by the rolling American Community Survey (ACS), which samples approximately 3.54 million addresses per year. To ensure that its estimates are reliable and of high quality, ACS data are often reported as *period* as opposed to *point* estimates. That is, for small geographic units such as census tracts, ACS estimates are reported for five years' worth of data. The Brown University LTDB that has been used throughout this chapter contains selected ACS data, at the census tract level, for the 2006–2010 five-year vintage of the survey. Among the ACS variables contained in the dataset is *per capita income* (PCI). PCI is the total dollar amount of income reported in a geographic unit, divided by the population of that unit. PCI can therefore be multiplied by the total population of a geographic unit to derive a measure of that unit's *aggregate income* (in U.S. dollars).

Regardless of which variable is the first to change, it is generally believed that a shrinking population is directly correlated with a shrinking economy or wealth base (Schilling and Mallach 2012). As will be expanded upon in later chapters, urban shrinkage is often connected to a process of household mobility known as **filtering**. In short, relatively wealthy households relocate from depreciating housing units to newer units, thereby making their depreciated homes (in terms of both physical quality and economic value) available to lower-income households. The implication is that older neighborhoods are eventually "succeeded by" groups with lower average income relative to the neighborhood's erstwhile occupants (Grigsby et al. 1987). Together, these observations lead to the following two additional research questions:

6 Did per capita income (i.e., average income per person) decrease in tracts that experienced population shrinkage from 1970 to 2010?
7 Did tracts that experienced population shrinkage from 1970 to 2010 also shrink in aggregate income (and vice versa)? In other words, is *population shrinkage* associated with *shrinkage in economic wealth* (i.e., income)?

Of the 52,380 conterminous census tracts included in the Brown University LTDB, 52,105 (99.5 percent) contain valid entries for PCI in both 1970 and 2010. However, prior to using these variables to investigate research questions #6–7, it is necessary to ensure that the nominal income data from the 1970 census long form survey are compatible with the nominal income data from the 2006–2010 ACS. Stated differently, the variable PCI in 1970 from the Brown University LTDB must be adjusted for inflation. Hence, in consultation with the inflation calculator provided by the U.S. Bureau of Labor Statistics (BLS), the 1970 PCI data are inflated (i.e., multiplied) by 5.62 to represent their equivalent value in 2010 dollars (BLS n.d.).

With comparable measures of PCI for 1970 and 2010 in hand, question #6 can be addressed using a paired (related) sample statistical test on the set of shrinking tracts that were identified in Figure 2.2. Among the 7,386 shrinking tracts pictured in Figure 2.2, 7,226 (97.8 percent) contain valid entries for both of the PCI variables. As with virtually all income data in the United States, the distributions of these variables are extremely positively skewed (skewness equals 4.17 for inflation-adjusted PCI in 1970, and 3.34 for nominal PCI in 2010). For that reason, a non-parametric Wilcoxon rank-sum test, which effectively compares the medians of two related samples, is used to evaluate whether PCI decreased in shrinking tracts from 1970 to 2010. Table 2.5 displays the results from this test. From 1970 to 2010, median PCI in tracts that experienced population shrinkage (a minimum of 25 percent population loss over the four decades) declined by $2,333.92. The difference is highly statistically significant ($p << 0.001$), which supports the notion that population shrinkage plausibly accompanies – or is accompanied by – the type of residential *filtering* described above. Moreover, Table 2.5 also shows the results of a parallel related samples Wilcoxon test for the non-shrinking tracts in the study area (n = 44,879 valid observations). During the same time period that PCI significantly decreased in shrinking tracts, the median PCI in non-shrinking tracts experienced a statistically significant *increase* of $1,417.20.

Next, regarding research question #7, PCI in 1970 (inflation adjusted) and 2010 are multiplied by their corresponding total population values to create measures of *aggregate income* for each of the two time periods. These derived variables are then used to compute the percentage change in aggregate income for all census tracts for the four-decade period from 1970 to 2010. Adopting the same threshold for shrinkage that was used in the earlier population analysis (i.e., a 25 percent or greater loss over 40 years), all tracts are coded as "shrinking" or "non-shrinking" with respect to their economic wealth. A useful tool for detecting a relationship between the statuses of *population shrinkage* and this latter form of *economic shrinkage* is a contingency table. In Table 2.6, rows classify tracts according to their population shrinkage status, and columns do the same for economic shrinkage status. The cell frequencies allow one to tally the number of tracts that experienced both population *and* economic shrinkage. As shown in the table, 5,715 of 52,105 tracts fall into this "shrinking-shrinking" category. If there were no association

Table 2.5 Wilcoxon related samples rank-sum test for equality of medians by tract type

Population Status	*Median Per Capita Income, 1970 (2010$)*	*Median Per Capita Income, 2010 (2010$)*	*Difference (2010$)*	*n*
Shrinking	21,191.42	18,857.50	–2,333.92***	7,226
Non-Shrinking	25,326.80	26,744.00	+1,417.20***	44,879

***$p << 0.001$

between population shrinkage and economic shrinkage, then one should expect this number to be approximately 1,393. That is, the probability that a tract experienced population shrinkage in the overall sample is 7226/52105 = 0.1387, and the probability of economic shrinkage is 10041/52105 = 0.1927; so, the expected number of tracts that experienced *both* population and economic shrinkage is [(0.1387 * 0.1927) * 52105] = 1393.

Clearly, there is a large discrepancy between the observed and expected number of tracts in this category. When expected and observed values are compared over all four table cells in this manner, a Pearson's chi-squared statistic can be computed to test the null hypothesis that there is no relationship between population and economic shrinkage. This null hypothesis is easily rejected ($p << 0.001$), which implies that population and economic shrinkage are dependent: knowing that a tract experienced population (economic) shrinkage improves one's ability to predict whether the tract also experienced economic (population) shrinkage.

Lambda (λ) is a common measure of association used to quantify the degree to which knowledge of one variable improves one's ability to predict the value of a second variable (Agresti and Kateri 2011). Lambda is an asymmetric measure of association that ranges from 0 (no improvement) to 1 (variable 1 can be perfectly predicted given knowledge of variable 2). The fact that λ is asymmetric means that one variable is always designated as "dependent" and the other as "independent". Table 2.6 indicates that λ equals 0.192 when *population shrinkage* is the dependent variable, and 0.419 when *economic shrinkage* is the dependent variable. In plainer terms, knowing that a tract experienced economic shrinkage (independent variable) decreases the number of errors one would make in attempting to predict cases of population shrinkage (dependent variable) by approximately 19 percent. Similarly, knowing that a tract experienced population shrinkage (independent variable) decreases the number of errors one would make in attempting to predict cases of economic shrinkage (dependent variable) by approximately 42 percent. The magnitude of this latter value is quite impressive, and potentially suggests that population shrinkage is more predictive of economic shrinkage than the converse. Nevertheless, regardless of which change occurs first, the results derived in this subsection demonstrate rather unequivocally that population shrinkage is associated with reductions in median PCI (question #6), as well as shrinkage in overall income (question #7).

Table 2.6 Analyzing the relationship between population shrinkage and economic shrinkage

	Shrinking Aggregate Income	*Non-Shrinking Aggregate Income*	*Total*
Shrinking Population	5,715	1,511	7,226
Non-Shrinking Population	4,326	40,553	44,879
Total	10,041	42,064	52,105
λ (row variable is dependent):	0.192		
λ (column variable is dependent):	0.419		

Pearson's chi-squared test: χ^2 [1] = 19.3×10^3; $p << 0.001$

Housing units (and occupied housing units) per acre

Drawing on the context laid out in the introduction to this chapter, the results of the preceding subsection paint a picture of a slow *negative feedback*, operating through the mediator variable income, which contributes to long-term population shrinkage in selected census tracts. Namely, population shrinkage was found to be strongly associated with economic (wealth) shrinkage, such that patterns of out-migration in shrinking areas leave behind populations characterized by lower average economic status relative to the pre-shrinkage population. The implication is that areas affected by sustained population loss tend to have fewer financial resources with which to maintain residential structures, relative to the stock of resources available prior to the onset of shrinkage. An expected outcome of this situation is a lower average level of residential property maintenance (Galster, Cutsinger, and Malega 2006). This outcome, in turn, can make a shrinking neighborhood a physically less attractive place to live. One response to a downgrade in a neighborhood's physical appearance is for more households to relocate (Bourne 1981). Thus, economic shrinkage can create a feedback effect that reinforces patterns of population shrinkage (Hospers 2014).

An additional variable subject to shrinkage, and which is directly implicated in the feedback processes described above, is therefore the residential built environment. Economists Edward Glaeser and Joseph Gyourko (2006) cite the durability of housing as one of the most important factors involved in the ongoing production of urban shrinkage and, ultimately, urban decline (Ch. 3). That is, because homes are typically "built to last" (and expensive to demolish), the pace of household out-migration is far faster than the rate at which housing units are removed from a shrinking place's urban fabric. Therefore, urban shrinkage is likely to produce a *mismatch* between the *scale* of an area's population and the *scale* of its built environment. Such situations are expected to result in high rates of vacancy and property abandonment, substantially lower property values, visual blight and disorder, and increasingly concentrated poverty (Glaeser and Gyourko 2006; Weaver and Bagchi-Sen 2013). While many of these phenomena are manifestations of urban *decline*, and are consequently discussed in the next chapter, the notion that the residential built environment can undergo [relatively slow] shrinkage leads to at least three additional research questions:

8 To what extent, if at all, did the scale of the residential built environment shrink in census tracts that experienced population shrinkage?
9 How did contraction of the residential built environment in shrinking tracts, if any, compare to the magnitude of population loss in those tracts?
10 What do the answers to questions #8–9 imply about residential land use change in shrinking tracts?

To address these questions, it is once again possible to rely on the "full count" (decennial census) data available in the Brown University LTDB. As stated earlier, data on the total number of [occupied] housing units are standard outputs of the decennial census. Of the 7,386 census tracts in the Brown University dataset that were identified as shrinking in Figure 2.2, 7,384 (99.97 percent) of them contain valid entries for housing units and occupied housing units in both the 1970 and 2010 decennial censuses. For this set of 7,384 tracts, Esri's ArcGIS® was used to compute each tract's physical area in acres. Following Hollander (2011), the number of housing units per acre ("housing unit density") is assumed to be a reliable indicator of the scale of the residential *built environment* of a census tract; and the number of occupied housing units per acre ("occupied unit density") is taken as a surrogate measure for the extent of residential *land use* in a census tract. When evaluated in tandem with the number of persons per acre ("population density") in a tract, changes in housing unit and occupied unit density between 1970 and 2010 should reveal critical information about the magnitude, relative pace, and land use implications of shrinkage in the residential built environment of shrinking census tracts.

Since housing unit, population, and occupied unit density are all count variables (adjusted for area), their distributions are positively skewed. For that reason, nonparametric Wilcoxon rank-sum test like those used in the previous subsection are performed here to evaluate median changes in the three density variables from 1970 to 2010. Table 2.7 contains the results from these tests. First, from 1970 to 2010, median housing unit density in tracts that experienced population shrinkage decreased by −0.58 units per acre. The difference is highly significant ($p << 0.001$), meaning that shrinking tracts did in fact experience contraction in their [median] residential built environment scales (question #8).

Also, median population density in shrinking tracts decreased by approximately −4.40 persons per acre over the course of the four most recent decennial censuses ($p << 0.001$). In other words, the observed change in median population density was roughly 7.6 times greater in magnitude than the observed change in median housing unit density. Note that the median number of persons per housing unit in the sample of shrinking tracts was 3.04 in 1970. Hence, if the median change in the residential built environment was somewhat proportional to the median change in population, then one might expect median population density to decrease at around three times the rate of housing unit density. In fact, though, median population density fell nearly eight times faster than median unit density (question #9), thereby supporting the thesis that a place's built environment is slow to adjust to much faster-moving changes in its population (Glaeser and Gyourko 2006).

The implications of these results for residential land use are stark. Namely, the observed change in median *occupied* unit density was approximately −0.85 units per acre about one and a half times the magnitude of the observed change in median

Table 2.7 Wilcoxon related samples rank-sum tests for equality of medians in various density measures

Variable	*Median in 1970*	*Median in 2010*	*Difference*
Housing Unit Density (units per acre)	3.48	2.90	–0.58***
Population Density (persons per acre)	10.55	6.15	–4.40***
Occupied Unit Density (units per acre)	3.32	2.47	–0.85***

n = 7,384 of 7,386 shrinking census tracts (see Fig. 2.2); ***p << 0.001

overall housing unit density. That both median population density and median occupied unit density seemed to decrease in shrinking tracts far faster than median total housing unit density suggests that population shrinkage does in fact create a *scale mismatch*: following periods of long-term population shrinkage, the rapidly falling number of residents in a given place is insufficient to occupy the place's only marginally decreasing number of housing units (question #10). The expected results of this mismatch include systemic issues of residential property disinvestment, vacancy, and abandonment (Schilling and Mallach 2012). Because these issues are generally considered to be indicators of *urban decline*, they are skipped over here and picked up in the next chapter.

Concluding remarks

This chapter set out to describe patterns and trends in urban shrinkage in the United States, with a particular emphasis on changes that have occurred at the census tract level over the past four decades. Two overarching objectives were pursued throughout the chapter. First, readers were introduced to a variety of data sources, indicators, analytical tools, and empirical approaches for studying urban shrinkage in the United States. This chapter combined the threshold method of identifying shrinking areas with time-varying population shrinkage centroids and location quotients to explore current and projected patterns of tract-level urban population shrinkage in the United States. The chapter adds to the suite of tools currently available for analyzing shrinkage with quantitative data. Second, the chapter posed and attempted to answer ten inter-related questions that probed the multiple dimensions in which – or stock variables on which – shrinkage operates. The ten questions are repeated in Table 2.8 for convenience.

Table 2.8 Questions addressed in Chapter 2

1. Where are the census tracts that meet the adopted threshold for shrinkage located?
2. Where are census tracts that appear to be on target to meet the adopted 40-year threshold for shrinkage over the next 10, 20, and 30 years located?
3. Where are the census *places* that meet the adopted threshold for shrinkage located?
4. Are there growing (shrinking) tracts within shrinking (growing) places?
5. Are shrinking census tracts found disproportionately in shrinking places? Does this pattern appear to be changing with time?
6. Did per capita income (i.e., average income per person) decrease in tracts that experienced population shrinkage from 1970 to 2010?
7. Did tracts that experienced population shrinkage from 1970 to 2010 also shrink in aggregate income (and vice versa)? In other words, is *population shrinkage* associated with *shrinkage in economic* wealth?
8. To what extent, if at all, did the scale of the residential built environment shrink in census tracts that experienced population shrinkage?
9. How did contraction of the residential built environment in shrinking tracts, if any, compare to the magnitude of population loss in those tracts?
10. What do the answers to questions #8–9 imply about residential land use change in shrinking tracts?

The tentative answers to these ten questions that were derived throughout this chapter offer the following five key ideas or takeaway points:

- **The most prominent and widely studied form of urban shrinkage is *population* shrinkage**. While at least two different techniques are used by researchers to identify shrinking places based on population loss, it is generally agreed that urban shrinkage involves *severe*, *persistent*, and *prevalent* decreases in the total number of persons living in an affected area (Beauregard 2003, 2009; Schilling and Logan 2008; Hollander 2011; Schilling and Mallach 2012).
- ***Actually existing geographies* of urban population shrinkage are surprisingly dispersed**. That is, population shrinkage does not apply exclusively to older industrial cities in the Midwestern and Northeastern parts of the United States. To the contrary, shrinking tracts are found all throughout the country (Fig. 2.2), including in "non-shrinking" cities (Table 2.4) in the south and west (Fig. 2.2). What is more, the prevalence of shrinking census tracts is on the rise in the "Sun Belt", while it is becoming somewhat stabilized in the "Rust Belt" (Fig. 2.4).
- **Still, the distribution of shrinking *cities* remains highly concentrated in the "Rust Belt"**. Moreover, current rates of population change suggest that this clustered pattern of shrinking places is not likely to change significantly in the near future (Fig. 2.5).
- **Population shrinkage is not independent of economic shrinkage (and vice versa)**. Population shrinkage is strongly and statistically significantly associated with shrinkage in the per capita and aggregate income of a given census tract (Tables 2.5–2.6). When population shrinks, the wealth of the shrinking area is prone to shrinkage as well. In cases where economic shrinkage (a mediator variable) lowers the quality of and demand for residential dwellings in a given census tract, further out-migration is a likely outcome (Hospers 2014). This slow-moving *negative feedback* is therefore what Chapter 1 called a "downward spiral": population shrinkage begets economic shrinkage, which begets additional population shrinkage, and so on (Hollander 2011).
- **Population shrinkage occurs disproportionately faster than shrinkage in the scale of the residential built environment**. Residential housing is generally built to last. When residents move out of a shrinking location, they leave behind durable housing units, many of which remain empty for long periods of time (Glaeser and Gyourko 2006). In census tracts characterized by population shrinkage, the magnitude of population de-densification has far exceeded the rate at which empty structures can be removed from the residential built environment (Table 2.7). The result is usefully described as a *scale mismatch*. Simply put, there are too few residents to occupy the number of available housing units. As a consequence, the number of *occupied* housing units per acre has decreased significantly in shrinking census tracts (Table 2.7). The implication is that the number of vacant and abandoned – and thus potentially deteriorating – housing

units increases as a place experiences population shrinkage. Such an outcome constitutes a qualitative change to a place's urban fabric that has the capacity to break up the place's existing urban form (Mumford 1961; Hollander 2011). In other words, *urban shrinkage* is seemingly inextricably linked to *urban decline* (Ch. 1). Hence, the focus will now shift (in Chapter 3) to analyzing patterns of and trends in urban decline in the United States.

Notes

1 https://www.census.gov/geo/reference/gtc/gtc_place.html
2 As tracts are the foundational unit used throughout the analysis, places were constructed from aggregate tract-level data from the Brown University LTDB.
3 Strictly speaking, income is a flow variable (i.e., a variable that contributes to a stock at a rate per unit time), not a stock. However, existing national data sources do not presently offer better measures of wealth that are readily accessible for multiple time periods and levels of geography (see Box 2.1). As such, income is adopted as a proxy for wealth in this part of the analysis.

References

Agresti, Alan, and Maria Kateri. 2011. "Categorical Data Analysis." In *International Encyclopedia of Statistical Science*, edited by Miodrag Lovric. Berlin: Springer.
Batty, Michael. 2013. "A Theory of City Size." *Science* 340 (6139):1418–1419.
Beauregard, Robert A. 2003. "Aberrant Cities: Urban Population Loss in the United States, 1820–1930." *Urban Geography* 24 (8):672–690.
Beauregard, Robert A. 2009. "Urban Population Loss in Historical Perspective: United States, 1820–2000." *Environment and Planning A* 41 (3):514–528.
Bourne, Larry S. 1981. *The Geography of Housing*. Toronto: V.H. Winston & Sons.
Bullock III, Charles S. 2010. *Redistricting: The Most Political Activity in America*. Lanham, MD: Rowman & Littlefield Publishers, Inc.
Flora, Cornelia Butler, Jan L. Flora, and Stephen Gasteyer. 2015. *Rural Communities: Legacy + Change*. 5th ed. Boulder: Westview Press.
Galster, George C., Jackie M. Cutsinger, and Ron Malega. 2006. "The Social Costs of Concentrated Poverty: Externalities to Neighboring Households and Property Owners and the Dynamics of Decline." In *The Future of Rental Housing*. Cambridge, MA.
Glaeser, Edward L., and Joseph Gyourko. 2006. "Housing Dynamics." In *Working Paper Series: National Bureau of Economic Research*. NBER Working Paper 12787, National Bureau of Economic Research, Cambridge, MA.
Grigsby, William, Morton Baratz, George Galster, and Duncan Maclennan. 1987. "The Dynamic of Neighborhood Change and Decline." *Progress in Planning* 28:1.
Hollander, Justin B. 2011. "Can a City Successfully Shrink? Evidence from Survey Data on Neighborhood Quality." *Urban Affairs Review* 47 (1):129–141.
Hospers, Gert-Jan. 2014. "Urban Shrinkage in the EU." In *Shrinking Cities: A Global Perspective*, edited by Harry W. Richardson and Chang Woon Nam. London: Routledge.
Kinevan, Marcos E. 1950. "Alaska and Hawaii: From Territoriality to Statehood." *California Law Review* 38 (2):273–292.
Kodrzycki, Yolanda K., and Ana Patricia Muñoz. 2015. "Economic Distress and Resurgence in U.S. Central Cities: Concepts, Causes, and Policy Levers." *Economic Development Quarterly* 29 (2):113–134.

Kwan, Mei-Po. 2012. "The Uncertain Geographic Context Problem." *Annals of the Association of American Geographers* 102 (5):958–968.

Logan, John R., Zengwang Xu, and Brian J. Stults. 2014. "Interpolating U.S. Decennial Census Tract Data from as Early as 1970 to 2010: A Longitudinal Tract Database." *The Professional Geographer* 66 (3):412–420.

MacDonald, Heather, and Alan Peters. 2011. *Urban Policy and the Census*. Redlands: Esri Press.

Mumford, Lewis. 1961. *The City in History*. New York: Harcourt, Brace & World.

Plane, David A., and Peter A. Rogerson. 1994. *The Geographical Analysis of Population with Applications to Planning and Business*. New York, NY: John Wiley & Sons.

Pumain, Denise. 2006. "Alternative Explanations of Hierarchical Differentiation in Urban Systems." In *Hierarchy in Natural and Social Sciences*, edited by Denise Pumain. Netherlands: Springer.

Schilling, Joseph, and Jonathan Logan. 2008. "Greening the Rust Belt: A Green Infrastructure Model for Right Sizing America's Shrinking Cities." *Journal of the American Planning Association* 74 (4):451–466.

Schilling, Joseph M., and Alan Mallach. 2012. *Cities in Transition: A Guide for Practicing Planners (Planning Advisory Service No. 568)*. Chicago: American Planning Association.

Turchin, Peter. 2003. *Complex Population Dynamics: A Theoretical/Empirical Synthesis*. Vol. 35, *Monographs in Population Biology*. Princeton: Princeton University Press.

Weaver, Russell C., and Sharmistha Bagchi-Sen. 2013. "Spatial Analysis of Urban Decline: The Geography of Blight." *Applied Geography* 40:61–70.

Weaver, Russell C., and Jason Knight. 2014. "Evolutionary Mismatch as a General Framework for Land Use Policy and Politics." *Land* 3 (2):504–523.

Yanow, Dvora. 2003. *Constructing "Race" and "Ethnicity" in America: Category-Making in Public Policy and Administration*. Armonk: ME Sharpe.

3 Patterns and trends

Measuring and mapping urban decline

Urban *decline* is often perceived to be synonymous with urban *shrinkage* (Dewar and Thomas 2012; Ryan 2012). As the evidence from Chapter 2 demonstrates, severe and persistent population loss is often strongly associated with reductions in an area's wealth base; and it typically creates a dysfunctional mismatch between the size of the affected area's population and the area's built environment. Together, these observations imply that as a place shrinks, it becomes ever more vulnerable to issues of concentrated poverty, neighborhood disinvestment, property abandonment, and chronic vacancy. These changes, in turn, can reinforce existing patterns of change or shrinkage as part of a "downward spiral" (Glaeser and Gyourko 2006; Hospers 2014).

Recall from Chapter 1 that shrinkage is not a necessary condition for decline. Decline can manifest in all varieties of settlements, regardless of whether they are growing, stable, or shrinking. For instance, a 2014 report published by the Brookings Institution found that the population of Charlotte, North Carolina, increased by approximately 30 percent from 2000 through 2012. During that same time period, the population of "poor" persons in the city increased by 98 percent. The number of such persons living in "high poverty" census tracts – where the report defined "high poverty" tracts as those where 40 percent or more of the total population lives below the federal poverty level – increased by a staggering 640 percent (Kneebone 2014). In plainer terms, distressed census tracts within a fast-growing city ostensibly declined. Poverty became substantially more concentrated in Charlotte, such that the city's "high poverty" areas became more disadvantaged or were worse off than they had been at the beginning of the study period (Kneebone 2014).

If both growth and shrinkage are capable of producing declining neighborhoods, then why does decline often appear synonymously with shrinkage in the urban literature? One reason could be that shrinkage is arguably a sufficient condition for decline in the near term. Namely, the fast pace of depopulation in shrinking places usually guarantees that such places cannot quickly adapt to their changing circumstances – and this lack of ability to adapt can lead to a downward spiral. For example, insofar as local government finances tend to rely heavily on property tax revenues, massive population (i.e., taxpayer) loss greatly reduces municipal operating budgets. It follows that shrinkage, at least in the short term, lowers the quality and quantity of goods and services that cities are able to provide to their residents (Mallach 2010). Additionally, population shrinkage translates into fewer "eyes on the streets", which may increase the likelihood of property and neighborhood disinvestment (Jacobs 1960). These ideas are well established in urban scholarship, and thus they are referenced as part of the theoretical survey found in Chapter 4.

A second and more immediately relevant reason that decline and shrinkage are so closely intertwined in the literature is that, due to the mutually reinforcing feedback effects discussed above, decline might be more prevalent, persistent, and severe in shrinking – as opposed to stable or growing – areas. In this chapter, patterns and trends in both decline and the co-occurrence of shrinkage and decline are examined using the Brown University LTDB (see Box 2.1). The next section looks at two concepts that are related to decline – *distress* and *disadvantage* – and then uses those concepts to operationalize a definition of urban *decline* that can be measured at the census tract level.

Decline, distress, and disadvantage

Much like shrinkage, decline is a *relative* concept. In this sense, decline is not a static variable that can be measured or evaluated at a single point in time. Rather, it is an active phenomenon that must be detected over time to determine whether a given place becomes "weaker", and thus more vulnerable.

Two comparatively static concepts that relate to this idea of a place's vulnerability to adverse changes are *distress* and *disadvantage*. These two terms appear rather frequently in urban scholarship (Großmann et al. 2013), where they are typically given operational definitions to facilitate quantitative analysis (Kneebone 2014). In most cases, the operational definitions are indistinguishable, despite referring to "distress" (Kasarda 1993) in some cases, and "disadvantage" (Manturuk, Lindblad, and Quercia 2009) in others. One potential explanation for their similarity is that these quantifiable representations of distress and disadvantage rarely flow out of theoretical definitions that situate the concepts, together with decline, inside a broader conceptual framework. Here, we first define *distress* and *disadvantage*. Then, by linking the two concepts with decline, we present a replicable strategy for detecting decline in empirical investigations.

Distress, defined by the Merriam-Webster dictionary as "state of danger or desperate need" (2015), can be interpreted as the degree to which a community is vulnerable to detrimental changes. Along those lines, distress is necessarily a function of disadvantage, where a **disadvantage** is "a quality or circumstance that makes achievement [of a goal or desired state] unusually difficult" (Merriam-Webster 2015). Assume that a goal of all geographic entities is to avoid harmful changes or decline (i.e., not experiencing decline is a "desired state" to be "achieved"). By definition, the presence of a disadvantage weakens a place's ability to realize this goal. It follows that places with more disadvantages are in greater "danger" of declining relative to places with fewer disadvantages. Using the terminology defined here, the implication is that place-based disadvantages increase place-based distress.

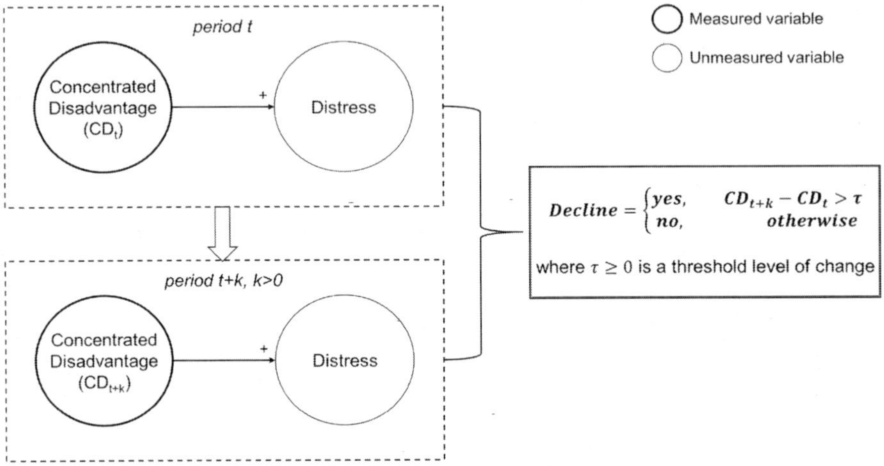

Figure 3.1 A framework for operationalizing "decline"

From this perspective, it is reasonable to conclude that a place is veritably distressed (i.e., in a state of desperate need, or in danger of decline) if it is characterized by *concentrated* disadvantage (Sampson, Morenoff, and Gannon-Rowley 2002). **Concentrated disadvantage** (CD) exists when multiple layers of disadvantage intersect in a single location. Several authors have proposed strategies for quantifying CD in the United States with indicators from Census Bureau datasets (Sampson, Morenoff, and Gannon-Rowley 2002; Manturuk, Lindblad, and Quercia 2009). These contributions inform our own development of a CD index using data from the Brown University LTDB in the next section. For more immediate purposes, the remaining task is to connect the concepts of CD, distress, and decline. Figure 3.1 illustrates how this can be done. The framework presented in Figure 3.1 is highly simplistic, resting on three key propositions.

First, as argued above, distress is an increasing function of concentrated disadvantage. This relationship is represented in Figure 3.1 by an arrow that extends from the latter concept to the former. The plus sign (+) near the end of the arrow signifies the positive nature of the relationship. In short, per the preceding definitions, while distress might be affected by a host of endogenous and exogenous sources (Adger and Brown 2009), endogenous increases in CD necessarily make a place more distressed. Second, comparable measurements of concentrated disadvantage can be observed at different points in time. At least in principle, the layering of multiple types of disadvantage in a single location can be documented with empirical data (Sampson, Morenoff, and Gannon-Rowley 2002). Finally, decline is detected via comparisons of these multiple, consistently measured, time-varying values of concentrated disadvantage. More precisely, the difference in CD between two time periods is a rough indicator of whether a place has become more distressed, or, in other words, **declined** over time.

Consistent with the two prominent approaches used to identify "shrinking" places that were presented in Chapter 2 – the binary and threshold methods – Figure 3.1 creates two broad possibilities for classifying places as "declining". First, under the **binary method**, any geographic unit for which CD increased between two time periods (t and $t + k$) are categorized as declining. Under this approach, the parameter τ in Figure 3.1 is set to zero, such that decline is detected whenever $CD_{t+k} > CD_t$. Alternatively, under the **threshold method**, τ is set to some critical value greater than zero such that only those geographic units that experienced increases in CD beyond τ are classified as declining. As we did in Chapter 2 with urban shrinkage, we adopt the latter of these methods for much of the remaining analysis.

Measuring concentrated disadvantage with U.S. census data

Disadvantage comes in many varieties, and it is impossible to conceive of, let alone enumerate, all of the qualities or circumstances that weaken a place's ability to avoid decline. Consequently, any measure of *concentrated disadvantage* (CD) is imperfect. With that limitation in mind, social science researchers generally concede that CD is measured for geographically-based populations. More accurately, social scientists have largely quantified CD based on attributes of a place's *people* rather than on attributes of the place itself (Sampson, Morenoff, and Gannon-Rowley 2002). There are many reasons for this choice, but at the heart of the approach lies the premise that certain population subgroups in the United States are, and historically have been, systematically disadvantaged in their collective abilities to succeed and achieve the same goals as other population subgroups (Sampson 2009). For now, Table 3.1 merely lists several of the variables that consistently feature in some of the most widely cited operational definitions of CD. The second column in the table indicates whether data for the given variable are available for all of the time periods (1970–2010) covered by the Brown University LTDB.

One possible method for measuring CD is to use the geometric mean of the first five variables listed in Table 3.1. (Recall that "Yes" in this column of the table indicates that the variables are included in the Brown University LTDB.) A **geometric mean** is a type of average calculated from the product, not the sum, of a set of values. A product-based average is appropriate when the mean is intended to capture a compound effect (Spizman and Weinstein 2008). Because *concentrated* disadvantage implies that multiple disadvantages coexist in the same place – and recognizing that, taken together, multiple forms of disadvantage tend to be self-reinforcing or *compounding* (Sampson, Morenoff, and Gannon-Rowley 2002) – a product-based mean is arguably both better equipped and more appropriate for quantifying CD than an additive approach. Furthermore, a geometric mean can be implemented for variables on different numeric scales and with different ranges of values (United Nations 2015). In our case, though, all variables are on the same scale. They are percentages that range in value from zero to 100. Therefore, the geometric average of the five available indicator variables listed in Table 3.1 can

Table 3.1 U.S. Census variables commonly used to measure concentrated disadvantage

Variable	*Available in the LTDB (1970–2010)?*
1. Percentage of the total population that is non-white	Yes
2. Percentage of households headed by a female with children present in the household	Yes
3. Percentage of persons 16 and over, in the civilian labor force, who are unemployed	Yes
4. Percentage of persons, for whom poverty status is determined, living below the federal poverty level	Yes
5. Percentage of persons aged 25 years or older with a high school degree or less	Yes
6. Percentage of households headed by a female	No (but see #2)
7. Percentage of households headed by single parents	No (but see #2)
8. Percentage of persons aged 16 to 19 not enrolled in school and not high school graduates	No
9. Percentage of the population receiving public assistance	No

Sources: Kasarda (1993); Sampson and Raudenbush (1999); Sampson, Morenoff, and Gannon-Rowley (2002); Manturuk, Lindblad, and Quercia (2009)

likewise range in value from zero to 100,[1] making the resultant measure of CD easy to interpret and consistently measurable across time periods. Finally, with a geometric mean, a change of *x* percent in one of the constituent variables has the same effect on CD regardless of which variable has changed (Spizman and Weinstein 2008).

That being said, the adopted **geometric index (*G-index*) of concentrated disadvantage** is computed as:

$$
\begin{aligned}
G_i = [&(Percent\ Nonwhite_i) \\
&* (Percent\ Female\ Headed\ Households_i)] \\
&* (Percent\ Unemployed_i) * (Percent\ in\ poverty_i) \\
&* (Percent\ Low\ Education_i)]^{\frac{1}{5}},\ i: 1, 2, \ldots, n
\end{aligned} \qquad [3.1]
$$

where *i* is an index that ranges over a set of *n* geographic units of analysis (e.g., census tracts), and the five variables on the right-hand side of the equation correspond to the first five variables listed in Table 3.1. Original data for these five variables, and all other variables enumerated in Table 3.1, are available through the U.S. Census SF3 (2000 and earlier) and the American Community Survey (Ch. 2). However, we rely on the Brown University LTDB to obtain data on the selected variables for 1970 and 2010 for current census tract boundaries.

The next section employs the G-index in an analysis of decline – and the relationship(s) between decline and shrinkage – for the same set of conterminous

U.S. census tracts (Fig. 2.1) that was used to study patterns of urban shrinkage in the preceding chapter.

Patterns of decline in the United States

Concentrated disadvantage, 1970 and 2010

Of the 52,380 conterminous census tracts in the Brown University LTDB (Logan, Xu, and Stults 2014) that contain interpolated data for current census tract boundaries going back to 1970 (Fig. 2.1), 51,784 (98.9 percent) feature valid entries for all five of the variables named in Equation 3.1 for both 1970 and 2010. This sample of 51,784 tracts allows us to analyze concentrated disadvantage (CD) over the same four-decade period for which we analyzed patterns of shrinkage in Chapter 2. The following three questions are answered below:

1 Was CD in 1970 greater in census tracts that were about to experience population shrinkage during the ensuing four decades relative to those that did not?
2 Was CD in 2010 greater in census tracts that experienced population shrinkage from 1970 through 2010 relative to those that did not?
3 Did the gap in CD between shrinking and non-shrinking tracts, if any, widen over the four-decade period of shrinkage (i.e., from 1970 to 2010)? In other words, did CD increase in shrinking tracts more rapidly than in non-shrinking tracts?

Similar to the analysis of income variables in Chapter 2, the time-varying distributions of the G-indices (Equation 3.1) in the sample dataset are highly skewed to the right. As a consequence, questions #1–2 are evaluated with nonparametric [independent samples] Wilcoxon signed-rank tests, which effectively compare the median difference in a variable between two groups. Table 3.2 presents the results of two such tests: one that compares the median G-index for shrinking tracts in 1970 to the median G-index for non-shrinking tracts in 1970 (question #1); and one that makes the same comparison for values of the G-index in 2010 (question #2). Recall that the G-index can range in value from zero to 100 (in the limit; see Appendix A), where higher values indicate greater/more severe CD.

Table 3.2 Wilcoxon independent samples signed-rank test for equality of medians

Census Year	*Median G-Index, Shrinking Tracts*	*Median G-Index, Non-shrinking Tracts*	*Difference*	*n*
1970	13.88	6.42	+7.46***	51,784
2010	25.20	10.92	+14.28***	51,784

***$p \ll 0.001$

Table 3.3 Comparison of a change in G between shrinking and non-shrinking tracts[a]

Tract Type	*Mean ΔG_i*[b]	*Standard Deviation*[c]	*n*
Shrinking	11.7	11.4	7,195
Non-Shrinking	8.3	8.9	44,589
Difference:(Shrinking – Non-Shrinking)	**+3.3*** **	0.14 (std. err.)	51,784

[a] Equal variances are not assumed; [b] The overall mean change for the combined sample is 8.9; [c] The overall standard deviation for the combined sample is 9.4; *** $p \ll 0.001$

Table 3.2 confirms that median CD, as measured by the G-index, was statistically significantly higher in tracts that experienced population shrinkage (Ch. 2) relative to non-shrinking tracts in both of the time periods under investigation. The implication of these results is that shrinking tracts appear to be more *distressed*, and thus more prone to *decline*, than non-shrinking tracts (Fig. 3.1). On that note, observe that the gap between the median G-index in shrinking and non-shrinking tracts widened over the forty-year period from 1970 to 2010 (question #3). In 1970, the median G-index in tracts that were about to experience four decades of persistent and severe population loss was 7.46 points higher than the median in all other census tracts. By 2010, this difference nearly doubled to 14.28 points. While median CD rose in all types of tracts over the given time frame (Table 3.2), the (small) increase was much greater in shrinking tracts relative to all other tracts.

To test the hypothesis that CD increased faster in shrinking tracts relative to non-shrinking tracts in a more formal way, we compute the forward difference in the G-index for each census tract in the sample as:

$$\Delta_{1970 \rightarrow 2010} G_i = \Delta G_i = G_{i,2010} - G_{i,1970},\ i: 1, 2, \ldots, n \qquad [2]$$

where the Greek letter Δ is read as "change in", G_i is the G-index for each census tract i in the set of 51,784 (n) census tracts, and the subscripts 1970 and 2010 refer to the starting and ending years under investigation, respectively.[2] Table 3.3 presents the results of a t-test to check if the average (mean) change in CD is equal in shrinking and non-shrinking tracts. The average ΔG_i in shrinking tracts exceeds the average ΔG_i in non-shrinking tracts by 3.3 points. In simpler terms, whereas CD increased on average in all types of census tracts, the increase observed in shrinking tracts was significantly greater (i.e., more severe) than in all other, non-shrinking tracts (question #3).

Detecting decline with the G-index

One of the broad questions posed at the outset of this chapter was whether there exists an association between decline and shrinkage. To assess whether there is an association between *shrinkage* and *decline*, it is necessary to operationalize the latter using either the *binary* or *threshold method*. Recall that the binary method advises us to classify a given census tract as having "declined" between 1970 and 2010 if CD in the tract increased during that forty-year interval (Fig. 3.1). One issue with this approach is implicated by the results of the t-test found in Table 3.3. The average tendency among all census tracts in the conterminous United States was for CD to move upward from 1970 to 2010. Indeed, CD – as measured by the G-index – increased in 45,442 (87.8 percent) of the 51,784 census tracts in the sample dataset. It is hard to justify a claim that nearly nine out of every ten census tracts in the United States declined over the past forty years. Rather, following the reasoning that led us to adopt the *threshold method* for identifying urban shrinkage in Chapter 2, we acknowledge here that the magnitude of change matters. If the increase in CD is not *severe* (i.e., at or beyond an adopted threshold), then a tract ought not to be classified as "declining".

Importantly, unlike shrinkage, for which planning experts and authorities on urban population loss have proposed quantitative thresholds for empirical analysis (e.g., Schilling and Logan 2008; Hollander 2011), there are no comparable established thresholds for the level of CD increase that corresponds to decline. Consequently, we adopt the following generic decision rule: *decline is an atypical increase in concentrated disadvantage*. Drawing on commonly used heuristics (Kasarda 1993), we define an "atypical increase" in CD as an increase in the G-index (from 1970 to 2010) of one or more standard deviations above the mean forty-year change. For a normally distributed variable, approximately 15.9 percent of all possible values lie one or more standard deviations above the mean. In our case, 15.7 percent of all tracts in the sample experienced a G-index increase that was one or more standard deviations above the mean change – which suggests that our adopted decision rule is useful for identifying "atypical increases".[3]

With these points in mind, the balance of this subsection is directed toward answering the following three additional research questions:

4 Where are the census tracts that meet the adopted threshold for decline located?
5 What is the geographic nature of the relationship between patterns of decline and patterns of shrinkage?
6 Is there an association between shrinkage and decline?

Concerning the first question (#4), Figure 3.2 maps the distribution of the 8,133 conterminous U.S. census tracts from the Brown University LTDB that experienced *decline* from 1970 through 2010. In other words, the tracts highlighted in Figure 3.2 witnessed atypical increases in CD, as measured by the G-index, over the past four decades. As was the case with urban *shrinkage* in Chapter 2, *declining* tracts are found all across the United States.

Notice in Figure 3.2 that declining tracts appear to be far less concentrated in the Middle Atlantic and North Central U.S. Census Divisions – which make up the majority of the American Rust Belt (Beauregard 2009) – compared to the distribution of shrinking tracts from Figure 2.3. To add empirical weight to this eyeball conclusion, a *location quotient* is computed for each of the nine large area

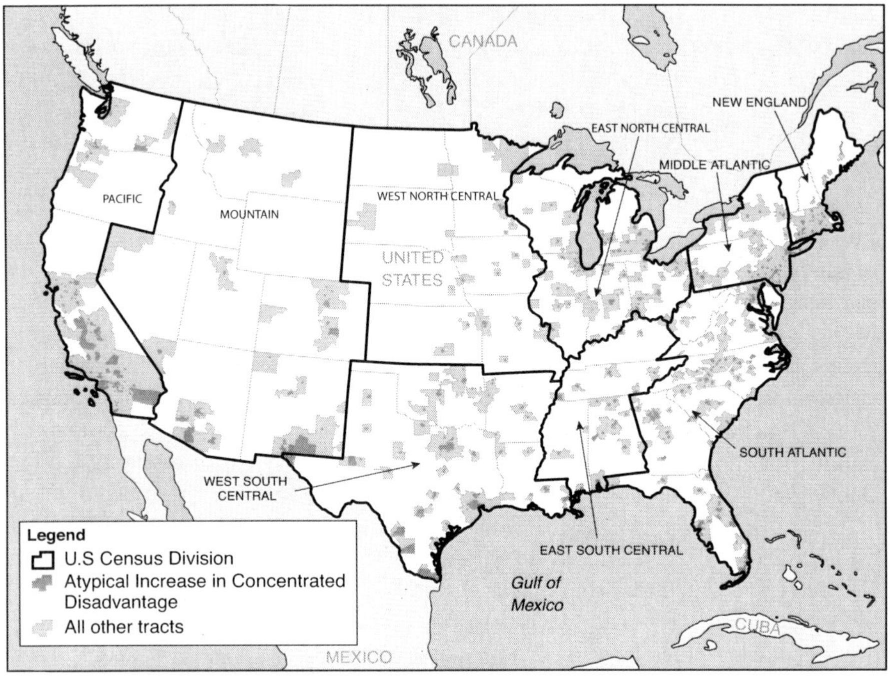

Figure 3.2 Distribution of census tracts in the conterminous United States that experienced decline from 1970 to 2010 (n = 8,133 declining tracts)

Census Divisions (Fig. 3.2) with respect to its share of declining census tracts. The procedure to calculate a Division-specific location quotient for declining tracts is the same as it was for shrinking tracts in the preceding chapter. Namely, for each Census Division, the location quotient for decline is given by:

$$W_i = \frac{d_i / n_i}{D / N}, \quad [3.3]$$

where W_i is the location quotient for Census Division i, d_i is the number of tracts classified as shrinking in Census Division i, n_i is the total number of tracts in Census Division i, D is the number of tracts classified as shrinking in the full dataset, and N is the total number of all tracts in the dataset. As before, values of 1 indicate that a given Census Division contains a proportionate share of the nation's declining tracts. Values greater than 1 indicate a relative concentration of decline, while values less than 1 describe disproportionately low shares of declining tracts.

Figure 3.3 plots the values of W_i for the nine U.S. Census Divisions. The graph shows that tract-level decline from 1970 through 2010 was relatively concentrated in the West South Central, East North Central, and East South Central Census

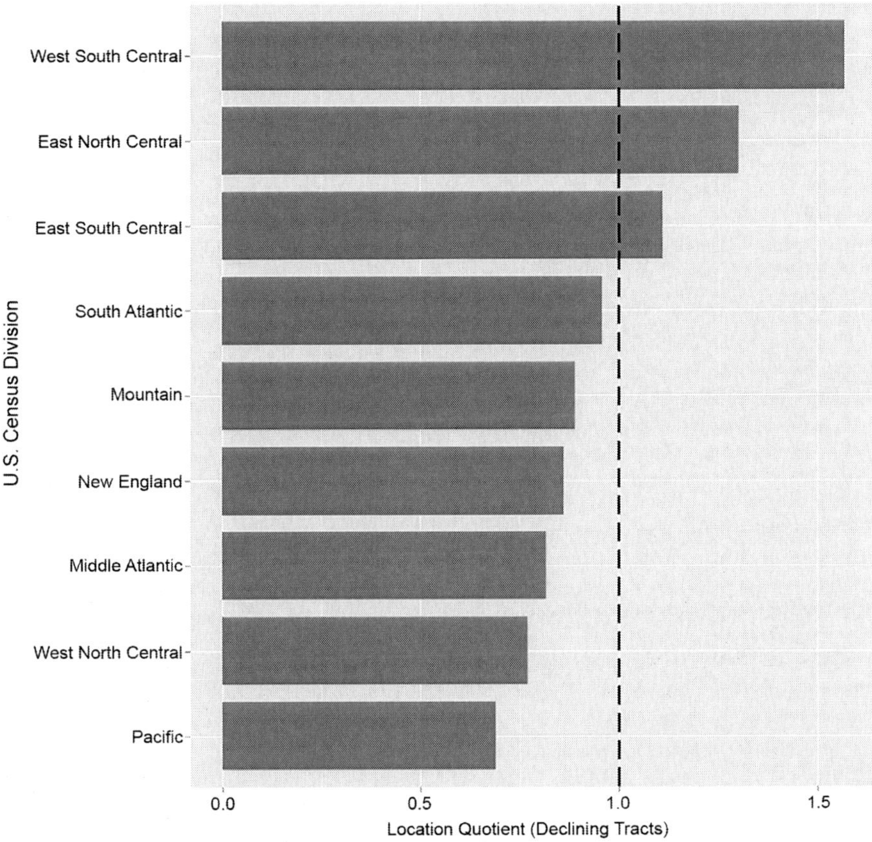

Figure 3.3 The concentration of declining tracts (1970–2010) in U.S. Census Divisions

Divisions (Fig. 3.2). Of these large geographic areas, only East North Central is commonly considered to be a part of the Rust Belt (Beauregard 2009). This Division is home to a disproportionate share of the nation's so-called "shrinking cities" (Fig. 2.5), and it is also associated with the greatest concentration of tract-level population shrinkage over the past four decades (Fig. 2.4). While there is an evident correlation between shrinkage and decline in this part of the Rust Belt, it is essential to point out that decline was most highly concentrated in the West South Central part of the *Sun* Belt (see Hollander 2011). Even though this Division contains the shrinking city of New Orleans, Louisiana (Table 2.3), it is predominantly made up – at least in the sample of tracts used in the analysis – of fast-growing Texas cities. The upshot is that, as stated earlier, shrinkage is not a necessary condition for decline (question #5). Decline operates in even the fastest-growing regions.

Table 3.4 Analyzing the relationship between population shrinkage and decline

	Declining Tracts	*All Other Tracts*	*Total*
Shrinking Tracts	1,849	5,346	7,195
All Other Tracts	6,284	38,305	44,589
Total	8,133	43,651	51,784
$\phi = 0.110$***			

Pearson's chi-squared test: $\chi^2[1] = 630.3$ ($p << 0.001$); ***$p << 0.001$

Still, just because decline and shrinkage do not follow identical geographical patterns does not mean that the two phenomena are not closely related. To the contrary, as argued many times in this book, shrinkage and decline tend to be mutually reinforcing (Glaeser and Gyourko 2006). Hence, regardless of the fact that both growing and shrinking places can decline, we hypothesize that decline is both more *prevalent* and more *severe* in shrinking areas relative to non-shrinking areas. The latter of these issues (severity) was studied above in Tables 3.2 and 3.3 and is taken up again in the next subsection. Here, a contingency table analysis (Table 3.4) aims to reveal whether decline is indeed more *prevalent* in shrinking census tracts compared to all other tracts. As a byproduct of this analysis, the magnitude and statistical significance of the association between shrinkage and decline in our sample of census tracts is quantified and interpreted (question #6).

The rows in Table 3.4 classify census tracts as "Shrinking" or "Other" based on the four-decade analysis of population change from Chapter 2 (see especially Fig. 2.2). The columns in the table divide the sample of tracts into those that experienced decline (Fig. 3.2) and those that did not, according to the analyses undertaken thus far in Chapter 3. The contingency table reveals that approximately 25.7 percent of tracts that experienced population shrinkage (1,849 of 7,195) also experienced atypical increases in CD (i.e., *decline*). Among all other, non-shrinking tracts, this figure was 14.1 percent. Interpreting the table in a slightly different way, while shrinking tracts constitute only 13.9 percent of the analytical sample, they account for 22.7 percent of the declining tracts. A chi-squared test of the null hypothesis that shrinkage and decline are independent of one another reveals that this disproportionality is highly statistically significant ($p << 0.001$). Stated another way, there is a direct association between shrinkage and decline. In the final row of Table 3.4, this association is quantified by the parameter phi (ϕ). ϕ is a symmetrical measure that, for a two-by-two matrix like Table 3.4, is tantamount to a correlation coefficient (with a maximum positive value of 1.0). In our sample, ϕ – again, a measure of the association between shrinkage and decline – is highly statistically significant, which supports our hypothesis and reinforces the results of the chi-squared test. At the same time, the relatively small magnitude of ϕ serves as a reminder that decline is not associated *exclusively* with shrinkage. Instead, as Figures 3.2 and 3.3 demonstrate, decline is a geographically dispersed phenomenon that affects places irrespective of the directions and degrees of their population changes.

The severity of decline in shrinking census tracts: indicators and evidence

There appears to be an interaction effect between shrinkage and decline, such that their co-occurrence makes places worse off than they would be if they were subjected to either shrinkage or decline alone, but not both. To test this assertion, we leverage results already derived in this book to classify census tracts from the Brown University LTDB into four groups: (1) tracts that experienced neither shrinkage nor decline from 1970 to 2010 ("Neither"); (2) tracts that experienced shrinkage, but not decline, from 1970 to 2010 ("Shrink only"); (3) tracts that experienced decline, but not shrinkage, from 1970 to 2010 ("Decline only"); and (4) tracts that experienced both shrinkage and decline from 1970 to 2010 ("Both"). Using this classification system, we are able to analyze group differences in various indicators to address two additional critical research questions:

7 Are the *people* in tracts that underwent both shrinkage and decline now more *distressed* than the *people* in tracts that experienced (a) one but not both, or (b) neither of these phenomena?
8 Are the *housing units* of tracts that underwent both shrinkage and decline now more *distressed* than the housing units of tracts that experienced (a) one but not both, or (b) neither of these phenomena?

Recall now that shrinkage was found to operate not only on an area's human population but also on its built environment (Table 2.7). For that reason, a clearer picture of the hypothesized interaction effect between shrinkage and decline might be obtained by examining group differences in various *physical* indicators that speak to the distress of an area's built structures. The U.S. Census American Community Survey (ACS) provides at least six serviceable indicators of physical distress (built environment vulnerability) in a given area. First, the ACS reports the total number of housing units in each census tract, along with the number of those units classified as "vacant". The fraction of all housing units that are vacant offers a proxy measure for vacancy in the given tract. It should be noted that **vacancy** refers to non-occupancy, that is, a vacant housing unit is not occupied by a full-time householder. Second, vacant units are broken down by type (e.g., vacant for rent, vacant for sale, seasonal, and other). In discussing the classification scheme used by the Census Bureau to assign a type to a vacant housing unit, Schilling and Logan (2008) argue that the category "Other Vacant" is a suitable surrogate measure of property abandonment. **Abandonment** occurs when an owner stops performing at least one essential responsibility of property ownership, where ownership responsibilities include, among others, maintenance, taxpaying, and occupancy (Mallach 2005). Compared to simple non-occupancy (vacancy), abandonment is more likely to result in visual blight and neighborhood decay (Schilling and Logan 2008). Third, the ACS reports the number of housing units constructed during ten-year intervals, where intervals start with a year ending in zero, beginning with 1940. That is, the ACS reports the number of units built before 1940, between 1940 and 1949, between 1950 and 1959, and on down the line. Because housing units are durable structures that depreciate in quality and value over time, the fraction of a place's housing stock that is considered to be "old" is an indicator of the amount of depreciation already present in the place's built environment. Whereas "old" is a subjective characteristic, it is generally agreed that old housing stock can be measured as the fraction of units that were built at least thirty years ago (Spatial Structures in the Social Sciences 2012).

The fourth through sixth physical distress indicators considered here are part of an ACS table called "Selected Conditions", and they are used to assess "the quality of [a place's] housing inventory". As such, planners and researchers use these indicators to measure the "vulnerability" (i.e., distress) of places' housing stocks at given points in time (Community Commons 2015: 6). While the Selected Conditions ACS table also features variables that deal with housing finances (i.e., *social* indicators), the three *physical* indicators of distress included in the table are:

- The count of housing units for which the number of occupants exceeds the number of rooms → this situation results in an occupant-to-room ratio greater than one, which is expected to increase wear and tear on a housing unit beyond that which comes from normal use;
- The count of housing units that lack complete plumbing facilities → complete plumbing facilities include (1) hot and cold running water, (2) a flush toilet, and (3) a bathtub or shower. If any one of these items is not present in a housing unit, then the unit is said to "lack complete plumbing facilities" (Community Commons 2015: 7); and
- The count of housing units that lack complete kitchen facilities → complete kitchen facilities include (1) a sink with a faucet, (2) a stove or range, and (3) a refrigerator. If any one of these items is not present in a housing unit, then the unit is said to "lack complete kitchen facilities" (Community Commons 2015: 7).

Table 3.5 summarizes the social and physical indicators that are drawn on in this subsection to address questions #7–8 from above. All data for the indicators listed in the table come from the most recent five-year vintage of the ACS (2010–2014), which was released in December 2015.[4]

Because the thirteen variables enumerated in Table 3.5 are all assumed to contribute to distress, the values of these indicators are expected to be highest in tracts that experienced both shrinkage *and* decline between 1970 and 2010. A corollary of this proposition is that the indicator values in these tracts will be statistically significantly higher than their corresponding values in tracts that experienced either shrinkage *or* decline, but not both. If, as we argue above, the co-occurrence of shrinkage and decline intensifies the severity of these phenomena, then this latter outcome is an observable implication of the interaction effect.

Table 3.5 Census variables used to analyze the interaction between shrinkage and decline

Social Indicators	*Physical Indicators*
Percentage of persons, for whom poverty status is determined, living below the federal poverty level (poverty)	Percent of housing units that are vacant (vacancy)
Percentage of persons aged 18 or younger, for whom poverty status is determined, living below the federal poverty level (childhood poverty)	Percent of housing units that are vacant and classified as "Other Vacant" (abandonment) [see Schilling and Logan (2008)]
Percentage of persons, for whom poverty status is determined, living at or below 50 percent of the federal poverty level (extreme poverty)	Percent of housing units that were built more than thirty years ago (old housing stock)
Percentage of families with children headed by single parents (single parents)	Percent of housing units with 1.01 or more occupants per room (high occupancy)
Percentage of households receiving public assistance income (public assistance)	Percent of housing units that lack complete plumbing facilities (lack of plumbing)
Percentage of persons 16 or older, in the civilian labor force, who are unemployed (unemployment)	Percent of housing units that lack complete kitchen facilities (lack of kitchen)
Percentage of persons aged 16 to 19 not enrolled in school and not high school graduates (school dropout)	

[a] Data for these variables were acquired through Social Explorer (Box 2.1) for the 2010–2014 five-year vintage of the U.S. Census American Community Survey (ACS), which was released in December 2015.

The Kruskal-Wallis (K-W) test (Tabachnick and Fidell 2013) can loosely be conceived of as a test for equality of medians between more than two groups. The rows in Table 3.6 show the results of one-way K-W tests that were executed on each of the indicator variables listed in Table 3.5. The low p-values reported in the fifth column of the table mean that the four groups of census tract – (1) Neither; (2) Shrink only; (3) Decline only; and (4) Both (see above) – exhibit significantly different median values for all of the selected social and physical contributors to distress that are named in Table 3.5.

While the p-values reported in Table 3.6 are telling – in that they reveal significant group differences in the medians of the indicator variables – each K-W test is an omnibus test. In other words, a K-W test allows us to conclude that groups are different; but not that one specific group is different from another specific group. Thus, if we wish to investigate the possibility that tracts in the "Both" category are now significantly worse off than tracts in the "Shrink only" and "Decline only" categories, then we need to engage in post hoc analysis. Post hoc analysis involves making pairwise comparisons between the different groups involved in a given K-W test. So, for example, a post hoc analysis of the K-W test for poverty, which is represented in the first row of Table 3.6, involves the following median

Table 3.6 Kruskal-Wallis (K-W) tests for equality of medians, by census tract status as shrinking, declining, neither, or both

	Neither	*Shrink Only*	*Decline Only*	*Both*	*K-W p-value*	*Evidence of Interaction*	*Stronger Tendency*
Social Indicators							
Poverty	9.7	18.6	29.3	**39.9**	<0.001	Yes	Decline
Childhood poverty	11.5	26.0	41.7	**56.1**	<0.001	Yes	Decline
Extreme poverty	4.0	7.9	11.8	**18.2**	<0.001	Yes	Decline
Single parents	11.2	17.8	26.8	**33.9**	<0.001	Yes	Decline
Public assistance	1.7	2.9	4.5	**7.0**	<0.001	Yes	Decline
Unemployment	7.6	10.2	13.8	**21.0**	<0.001	Yes	Decline
School dropout	0.0	0.0	5.3	**6.1**	<0.001	Ambiguous	Decline
Physical Indicators							
Vacancy	6.9	11.7	10.9	**21.1**	<0.001	Yes	Shrinkage
Abandonment	3.9	7.4	5.3	**14.6**	<0.001	Yes	Shrinkage
High occupancy	1.5	1.3	**7.1**	2.9	<0.001	Ambiguous	Decline
Lack of plumbing	0.0	1.8	1.3	**6.5**	<0.001	Yes	Shrinkage
Lack of kitchen	1.3	3.1	2.7	**9.3**	<0.001	Yes	Shrinkage
Old housing stock	61.5	88.7	79.0	**91.5**	<0.001	Yes	Shrinkage

Notes: The values reported in the table are median percentages (e.g., 9.7 in the first cell refers to 9.7 percent, which is the median poverty rate in sample tracts that experienced neither shrinkage nor decline). All possible pairwise comparisons in each row of the table are statistically significant at p<0.001 *except for* (1) the school dropout comparison between the two types of non-declining tracts, and (2) the same comparison for the two types of declining tracts. Post hoc pairwise tests were carried out with Conover's test for multiple comparisons of independent samples (Pohlert 2014) using a false discovery rate p-value correction method (Benjamini and Hochberg 1995). **Bold text** indicates the largest value in each row (i.e., the largest group median for the given variable). While one-way K-W tests were performed to facilitate the post hoc pairwise comparisons, an approximative multivariate K-W test (Johnson et al. 1993) was also carried out for all of the variables simultaneously using the *coin* package in R (Hothorn 2015). Consistent with the one-way p-values reported here, the p-value for the multivariate test was also <<0.001. Hence, there are meaningful *patterns* of differences in the group medians of the selected social and physical variables.

NB: n = 51,356 census tracts with valid entries for all thirteen variables in the 2010–2014 ACS

comparisons: (1) 9.7 percent poverty to 18.6 percent poverty; (2) 9.7 to 29.3; (3) 9.7 to 39.9; (4) 18.6 to 29.3; (5) 18.6 to 39.9; and (6) 29.3 to 39.9. The same logic holds for all other rows in the table. As the footnote on Table 3.6 discusses in greater detail, we carried out these pairwise analyses with Conover's test for multiple comparisons (Pohlert 2014) using a false discovery rate p-value correction technique (Benjamini and Hochberg 1995). With the exception of the school dropout variable, all pairwise comparisons revealed statistically significant different group medians at the 0.001 level of significance.

To make the foregoing results more interpretable in the context of our questions, Table 3.6 features two important modifications. First, bold text is used to highlight the largest group median for each indicator. In all but one case (the exception is high occupancy), these bold values appear in the "Both" column.

That is, of thirteen selected disadvantages that contribute to a census tract's *distress*, twelve are *most severe* in the tracts that concurrently shrank and declined from 1970–2010. Second, the column in Table 3.6 labeled "Evidence of Interaction" is used to summarize the results of two specific post hoc tests for each row. The two tests correspond to the pairwise comparisons between the median in the "Both" column and (1) the median in the "Shrink only" column and (2) the median in the "Decline only" column. If the post hoc tests found that the median for the "Both" group was significantly larger than the medians for both of the latter two groups, then the relevant row in the "Evidence of Interaction" column will say "Yes". In short, "Yes" in the Evidence column signifies that the co-occurrence of shrinkage and decline is associated with more severe distress relative to the singular occurrence of shrinkage or decline alone. Where the word "Ambiguous" appears in the evidence column, one of these two pairwise comparisons was significant and in the hypothesized direction, while the other was not. For school dropout, the median for the "Both" group was significantly larger than the median for the "Shrink only" group but not for the "Decline only" group. For high occupancy, the median for the "Both" group was significantly larger than the median for the "Shrink only" group; but it was significantly *less than* the median for the "Decline only" group.

The final column in Table 3.6 speaks to the pairwise post hoc tests that compare the medians for the "Shrink only" and "Decline only" groups. The point of this column is to identify the indicators for which shrinkage alone is tied to worse outcomes than decline alone, and vice versa. Hence, for any given row, if the median for the "Shrink only" group is significantly larger than the median for the "Decline only" group, the word "Shrinkage" appears in the final column. If the converse is true, then the word "Decline" appears in the last column. The findings show that social distress seems to be more of a function of decline than shrinkage; whereas physical distress seems to be more dependent on shrinkage than decline. The former of these results is most certainly an artifact of the approach we employed to operationalize decline. Recall that decline was operationally defined as an atypical increase in concentrated disadvantage (CD). Because CD is usually measured for human populations and not physical structures (Sampson, Morenoff, and Gannon-Rowley 2002; Manturuk, Lindblad, and Quercia 2009), declining tracts, by our definition (see Fig. 3.1), exhibit greater degrees of social distress relative to non-declining tracts. The latter result – *viz.*, the relatively high degree of physical distress in shrinking tracts – is interpreted here as further evidence of the *scale mismatch* that shrinkage creates between the size of a place's population and the size of its built environment. Namely, when an area experiences rapid, severe, and persistent population loss, the size of its [durable] built environment remains relatively static. The result is that there are substantially fewer people left to manage a mostly constant stock of housing units. As such, property disinvestment rises sharply, and the built environment of shrinking places depreciates at an accelerated pace (Schilling and Mallach 2012).

Collectively, the findings from Table 3.6 suggest that decline is most likely to amplify social distress and shrinkage is most likely to amplify physical distress in a given geographic area. However, when and where the two phenomena occur together, both physical and social distress become significantly more intense (severe). In other words, there is an interaction effect between shrinkage and decline, such that the two phenomena are mutually reinforcing.

Concluding remarks

As an extension of the previous chapter's analysis of urban *shrinkage*, this chapter attempted to describe patterns and trends in urban *decline* in the United States, again emphasizing changes that have taken place at the census tract level over the past four decades. In addition to introducing readers to various indicators, analytical tools, and empirical approaches for studying decline, Chapter 3 generated novel [tentative] answers to eight research questions that pertain to the association of – and interaction between – shrinkage and urban decline. These eight questions are repeated in Table 3.7 for convenience.

Chapter 3 investigated the questions in Table 3.7 with a series of analyses that were carried out on the Brown University LTDB (Logan, Xu, and Stults 2014; Box 2.1). Taken together, the results from these exercises offer at least four major takeaways:

- **The ability to empirically detect *decline* in a consistent manner across space hinges on the availability of longitudinal data**. By definition, decline involves change. As a consequence, decline cannot be quantified with variables that are measured at a single point in time (*cf.*, Downs 1999). Rather, evaluating whether a place "tended toward a weaker" or more vulnerable

Table 3.7 Questions addressed in Chapter 3

1. Was CD in 1970 greater in census tracts that were about to experience population shrinkage during the ensuing four decades relative to those that did not?
2. Was CD in 2010 greater in census tracts that experienced population shrinkage from 1970 through 2010 relative to those that did not?
3. Did the gap in CD between shrinking and non-shrinking tracts, if any, widen over the four-decade period of shrinkage (i.e., from 1970 to 2010)? In other words, did CD increase in shrinking tracts more rapidly than in non-shrinking tracts?
4. Where are the census tracts that meet the adopted threshold for decline located?
5. What is the geographic nature of the relationship between patterns of decline and patterns of shrinkage?
6. Is there an association between shrinkage and decline that cannot be explained by chance alone?
7. On average, are the *human populations* of tracts that underwent both shrinkage and decline now more *distressed* than the human populations of tracts that experienced (a) one but not both, or (b) neither of these phenomena?
8. On average, are the *housing unit populations* of tracts that underwent both shrinkage and decline now more *distressed* than the housing unit populations of tracts that experienced (a) one but not both, or (b) neither of these phenomena?

state calls for comparisons over time. In this sense, *decline* – an active/dynamic phenomenon – is plausibly reflected by an increase in *distress*. In turn, one surrogate/partial measure of distress is *concentrated disadvantage* (CD), or the coexistence of multiple attributes that diminish a place's ability to withstand decline (Fig. 3.1). The *G-index of concentrated disadvantage* developed in this chapter provides a means for quantifying CD with time-varying U.S. Census (socioeconomic) data that are available for a fixed set of geographic boundaries. As such, the G-index can be measured at multiple, static points in time, which gives researchers the ability to detect "atypical increases" in CD over time and space. Put differently, the G-index enables researchers to identify places that became more *distressed* between two given time periods.

- **Shrinking tracts exhibit higher concentrations of *disadvantage* relative to all other tracts**. Furthermore, the difference in CD between shrinking and non-shrinking tracts widens as population shrinkage progresses (Tables 3.2–3.3). In slightly more technical terms, there is an association between shrinkage and CD.
- **The geographic distribution of decline is dispersed, such that it affects all types of growing, shrinking, and stable settlements; however, decline is significantly more *prevalent* in shrinking tracts relative to all other tracts**. The actually existing geographies of decline are somewhat surprising. While there is a relative disproportion of declining tracts in the East North Central part of the Rust Belt (Fig. 3.2), where there is also a significant concentration of shrinking tracts (Fig. 2.4), declining tracts in our sample dataset are most heavily concentrated in the fast-growing West South Central part of the Sun Belt (Fig. 3.3). Yet, as Chapter 3 demonstrates, shrinkage is not a necessary condition for decline. That being said, regardless of these surprising macroscopoic patterns, at a finer- (census tract-) level of analysis, decline is indeed significantly more *prevalent* among shrinking tracts compared to all other tracts (Table 3.4).
- **In addition to being more *prevalent*, decline is also significantly more *severe* in shrinking tracts relative to all other tracts. More generally, the co-occurrence of shrinkage and decline is far more distress-inducing than the singular occurrence of shrinkage or decline alone**. Indicators of distress in both the people and the residential built environments of census tracts reveal strong evidence of an interaction effect between shrinkage and decline. Tracts in which both phenomena occur over the same time interval (here, from 1970–2010) appear to be substantially worse off than tracts where either shrinkage or decline occurred in isolation, or not at all.

Chapters 2 and 3 performed a variety of statistical analyses using a national longitudinal dataset (Logan, Xu, and Stults 2014) to reveal patterns and trends of urban shrinkage and decline in the United States over the past four decades. In the next two chapters (4–5), the book discusses some of the processes that are assumed to contribute to the production of these empirical patterns and trends. Two broad ideas inform our presentation: (1) shrinkage and decline are related but conceptually distinct phenomena that do not necessarily occur during the same spatiotemporal intervals or with the same degrees of force; but (2) where both phenomena are in fact active and appreciable, they tend to reinforce one another in ways that leave places in "desperate states of need". In other words, comparisons are needed to investigate some of the circumstances under which shrinking places have, at least for the moment, resisted sufficiently severe decline. Only then can we begin to identify what options are available for intervention to halt or reverse a downward spiral or negative cumulative causation. However, we claim no ability to successfully disentangle the extraordinary intricacies and complexities of distressed urban systems or develop policy solutions. Instead, from this point forward we seek only to provide readers with a primer on some of the contexts (e.g., place-specific) and processes that make escape from these adverse outcomes unusually difficult.

Notes

1 Additional details provided in Appendix A.

2 Unlike the distributions of the time-varying G-indices that were analyzed in Table 3.2, the distributions of the ΔG_i variables are bell-shaped and close to symmetrical (skewness = 0.60 for non-shrinking tracts, and 0.22 for shrinking tracts). While these distributions are considerably more peaked than a normal distribution (kurtosis = 1.04 and 1.43, respectively), their shapes allow us to evaluate differences in the mean of ΔG_i [between shrinking and non-shrinking tracts] with a t-test. For large samples, t-tests are robust to slight departures from the assumption of normally distributed data. When given a choice between a parametric t-test and a nonparametric Wilcoxon rank-sum test, the former tends to have greater statistical power (Verzani 2005).

3 NB: the overall mean of ΔG_i for the sample of 51,784 census tracts is 8.9, with a standard deviation of 9.4. Hence, the threshold used to detect decline in this chapter is $\tau = 18.3$ (Fig. 3.1). For future studies, we recommend performing a sensitivity analysis that operationalizes decline as a G-index increase in the range of 15 to 20 points over four decades.

4 Notice that the set of social indicators listed in Table 3.5 is more comprehensive than the set of indicators that was used to construct the G-index of CD earlier in this chapter. Because we are interested in creating a *current* snapshot of social distress in our sample of census tracts, we are no longer dependent on the variables that are presently included in the Brown University LTDB.

References

Adger, W. Neil, and Katrina Brown. 2009. *Vulnerability and Resilience to Environmental Change: Ecological and Social Perspectives*. Oxford: Blackwell Publishing Ltd.

Beauregard, Robert A. 2009. "Urban Population Loss in Historical Perspective: United States, 1820–2000." *Environment and Planning A* 41 (3):514–528.

Benjamini, Yoav, and Yosef Hochberg. 1995. "Controlling the False Discovery Rate: A Practical and Powerful Approach to Multiple Testing." *Journal of the Royal Statistical Society. Series B (Methodological)* 57 (1):289–300.

Community Commons. 2015. *Community Health Needs Assessment (CHNA) Boulder County Vulnerability Report*. Boulder: Community Commons.

Dewar, Margaret, and June Manning Thomas, eds. 2012. *The City after Abandonment*. Philadelphia, PA: University of Pennsylvania Press.

Downs, Anthony. 1999. "Some Realities about Sprawl and Urban Decline." *Housing Policy Debate* 10 (4):955–974.

Glaeser, Edward L., and Joseph Gyourko. 2006. "Housing Dynamics." In *Working Paper Series: National Bureau of Economic Research*. NBER Working Paper 12787, National Bureau of Economic Research, Cambridge, MA.

Großmann, Katrin, Marco Bontje, Annegret Haase, and Vlad Mykhnenko. 2013. "Shrinking Cities: Notes for the Further Research Agenda." *Cities* 35:221–225.

Hollander, Justin B. 2011. "Can a City Successfully Shrink? Evidence from Survey Data on Neighborhood Quality." *Urban Affairs Review* 47 (1):129–141.

Hospers, Gert-Jan. 2014. "Urban Shrinkage in the EU." In *Shrinking Cities: A Global Perspective*, edited by Harry W. Richardson and Chang Woon Nam. London: Routledge.

Hothorn. 2015. https://cran.r-project.org/web/packages/coin/coin.pdf

Jacobs, Jane. 1960. *The Death and Life of Great American Cities*. New York: Vintage Books.

Johnson, W. D., Mercante, D. E. and May, W. L. 1993. "A Computer Package for the Multivariate Nonparametric Rank Test in Completely Randomized Experimental Designs." *Computer Methods and Programs in Biomedicine* 40 (3), 217–225.

Kasarda, John D. 1993. "Inner-City Concentrated Poverty and Neighborhood Distress: 1970 to 1990." *Housing Policy Debate* 4 (3):253–302.

Kneebone, Elizabeth. 2014. *The Growth and Spread of Concentrated Poverty, 2000 to 2008–2012*. Washington, DC: Brooking Institution, Metropolitan Policy Program.

Logan, John R., Zengwang Xu, and Brian J. Stults. 2014. "Interpolating U.S. Decennial Census Tract Data from as Early as 1970 to 2010: A Longitudinal Tract Database." *The Professional Geographer* 66 (3):412–420.

Mallach, Alan. 2005. *Building a Better Urban Future: New Directions for Housing Policies in Weak Market Cities*. Denver, CO: Community Development Partnerships' Network.

Mallach, Alan. 2010. *Bringing Buildings Back: From Abandoned Properties to Community Assets: A Guidebook for Policymakers and Practitioners*. 2nd ed. Montclair: National Housing Institute.

Manturuk, Kim, Mark Lindblad, and Roberto G. Quercia. 2009. "Homeownership and Local Voting in Disadvantaged Urban Neighborhoods." *Cityscape* 11 (3):213–230.

Merriam-Webster. n.d. Springfield: Merriam-Webster. http://www.merriam-webster.com/dictionary/Springfield

Pohlert, Thorsten. 2014. The Pairwise Multiple Comparison of Mean Ranks Package (PMCMR). http://CRAN.R-project.org/package=PMCMR.

Ryan, Brent D. 2012. *Design after Decline: How America Rebuilds Shrinking Cities*. Philadelphia: University of Pennsylvania Press.

Sampson, Robert J. 2009. "Racial Stratification and the Durable Tangle of Neighborhood Inequality." *The ANNALS of the American Academy of Political and Social Science* 621 (1):260–280.

Sampson, Robert J., and Stephen W. Raudenbush. 1999. "Systematic Social Observation of Public Spaces: A New Look at Disorder in Urban Neighborhoods." *American Journal of Sociology* 105 (3):603–651.

Sampson, Robert J., Jeffrey D. Morenoff, and Thomas Gannon-Rowley. 2002. "Assessing 'Neighborhood Effects': Social Processes and New Directions in Research." *Annual Review of Sociology* 28:443–478.

Schilling, Joseph, and Jonathan Logan. 2008. "Greening the Rust Belt: A Green Infrastructure Model for Right Sizing America's Shrinking Cities." *Journal of the American Planning Association* 74 (4):451–466.

Schilling, Joseph M., and Alan Mallach. 2012. *Cities in Transition: A Guide for Practicing Planners*. Vol. 568, *Planning Advisory Service*. Chicago: American Planning Association.

Spatial Structures in the Social Sciences. 2012. Census Geography: Bridging Data from Prior Years to the 2010 Tract Boundaries. Brown University. http://www.s4.brown.edu/us2010/Researcher/Bridging.htm

Spizman, Lawrence, and Marc A. Weinstein. 2008. "A Note on Utilizing the Geometric Mean: When, Why and How the Forensic Economist Should Employ the Geometric Mean." *Journal of Legal Economics* 15:43–55.

Tabachnick, Barbara G., and Linda S. Fidell. 2013. *Using Multivariate Statistics*. Boston: Pearson.

United Nations. 2015. "Technical Notes." United Nations. http://hdr.undp.org/sites/default/files/hdr2015_technical_notes.pdf.

Verzani, John. 2005. *Using R for Introductory Statistics*. Boca Raton: Chapman & Hall/CRC Press.

4 Explanations of urban shrinkage and decline

Statistical analyses in Chapters 2 and 3 uncovered several trends that offer essential guidance for examining the processes that give rise to shrinkage and decline. Namely, actually existing shrinkage and actually existing decline are phenomena whose spatial distributions are not characterized by one-to-one correspondence (Table 3.4). As we saw in Chapter 3, not all places that experience population shrinkage also experience decline (as measured through atypical increases in concentrated disadvantage), and vice versa. In other words, if planners and policymakers are interested in the "downward spiral" in which some shrinking places find themselves (Hospers 2014), then understanding the differences between places that (1) shrank *and* declined and (2) shrank *but did not* decline, might reveal some general features of urban landscapes that help to break – or even forestall the development of – negative cumulative causation. Toward those ends, this chapter has four specific objectives.

First, we combine analytical findings from Chapters 2 and 3 to map the geographic distribution of U.S. census tracts that experienced both shrinkage *and* decline over the four-decade period covered by the Brown University Longitudinal Tract Database (LTDB; see Logan et al. 2014). Doing so is not intended to spark another lengthy round of data analysis but rather stimulate a discussion about the processes that might have produced patterns. The second objective of the chapter is to examine the role of three common processes underlying urban shrinkage and decline (deindustrialization, suburbanization, and demographic change) in relation to the patterns of shrinkage and decline. We argue that although these three factors are indispensable to understanding urban shrinkage and decline, they are not sufficient to explain why some shrinking places decline and others do not. Consequently, the third objective is to try to explain the patterns of coupled shrinkage and decline using existing urban theories. Finally, we support several of the claims made in this discussion through empirical analysis using the LTDB data.

Patterns of coupled shrinkage and decline in the United States

Drawing on findings from Chapter 2 (Fig. 2.2) and Chapter 3 (Fig. 3.2), Figure 4.1 maps the distribution of U.S. census tracts – for which four decades' worth of longitudinal data are available in the Brown University LTDB (Box 2.1) – that experienced both *shrinkage* and *decline* (measured as atypical increases in concentrated disadvantage) over the past forty years. Also pictured on the map are the geographic mean center of the tracts that both shrank and declined and the standard deviational ellipse associated with this set of tracts. The **geographic mean center** is the unique point derived by obtaining the arithmetic averages of the respective vectors of *x*- and *y*-coordinates for the highlighted tracts' centroids. The **standard deviational ellipse** (SDE) is created by computing the standard deviations of those *x*- and *y*-coordinates and using the resultant values to define the axes of an ellipse. The axes are then rotated at an angle that corresponds to the spatial orientation of the distribution (Lee and Wong 2001). Hence, the shape and area of the SDE convey valuable information about distributional trends in, as well as the geographic territory containing the majority of, the shrinking-declining tracts.

In Figure 4.1, nearly three-quarters (1,356 of 1,849) of all census tracts that experienced both shrinkage and decline between 1970 and 2010 are enclosed by the SDE (Note: the SDE is labeled "Core Area of Shrinkage and Decline" in Fig. 4.1). Moreover, the SDE is slightly elongated from northeast to southwest, extending from the center of the Middle Atlantic Census Division to the northern edge of the West South Central Division. These results seem to justify common

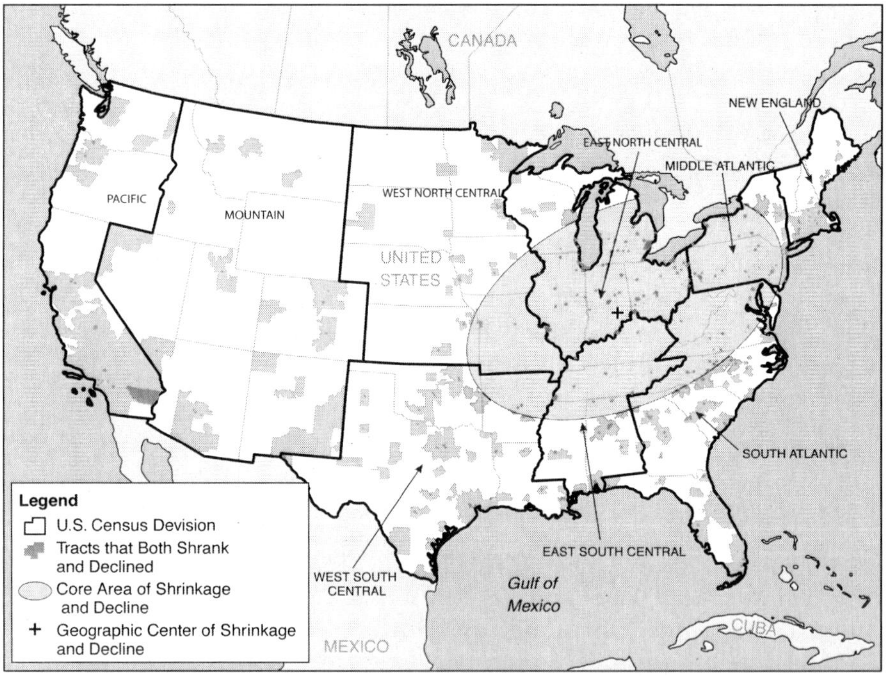

Figure 4.1 Distribution of census tracts in the conterminous United States that experienced both shrinkage and decline from 1970 to 2010 (n = 1,849 tracts)

perceptions that current manifestations of urban shrinkage and decline are largely Rust Belt (Beauregard 2009) phenomena that appear most frequently in older industrial cities in the Northeastern and Midwestern regions of the United States (Schilling and Mallach 2012). Thus, while it is certainly not the case that all places within the depicted core area endured shrinkage and/or decline during the temporal extent of our study, the co-occurrence of these two phenomena is far more prevalent in this area than in the rest of the country.

The preceding inference could probably be made with anecdotal evidence, without the need for rigorous data analysis. The image of a shrinking and declining Rust Belt has been inundating American popular culture and news media for decades. From the portrayal of severe distress in Baltimore, Maryland, in the HBO series *The Wire*,[1] to the MSNBC special feature *The Rust Belt: Once Mighty Cities in Decline*,[2] to national coverage of Detroit's historic bankruptcy,[3] older industrial cities in the Northeastern and Midwestern United States are well-known for their struggles with population loss, economic collapse, physical decay, and social inequality. In that vein, the pattern of coupled shrinkage and decline pictured in Figure 4.1 might not reveal sufficiently "new" information to students of urban shrinkage – but nor are they intended to do so. Rather, the point is that many participants in urban policy discourses focus on deindustrialization, suburbanization, and demographic change given the co-occurrence of shrinkage and decline specifically in the Rust Belt (Fig. 4.1). Put another way, because of their known effects on the Rust Belt (Beauregard 2006), deindustrialization, suburbanization, and demographic change are regularly proffered as the primary "causes" or "drivers" of shrinkage and decline (Großmann et al. 2013).

Deindustrialization, suburbanization, and demographic change

These three primary drivers (Schilling and Mallach 2012) of urban shrinkage are explored in somewhat greater detail in Chapter 5. For now, we offer only basic definitions to facilitate the remaining objectives of this chapter. First, **deindustrialization** is:

> [t]he rapid loss of factories and factory-related jobs to other parts of the country or to other nations or to the cessation or sharp reduction of production activities in key industries for any reason.
>
> (Bluestone, Huff Stevenson, and Williams 2008: 556)

Put differently, drawing on the definitions set forth in Chapter 1, **deindustrialization** refers to *shrinkage* in a place's manufacturing operations: it is a substantial, downward, quantitative adjustment to the total stock of employers and employees engaged in the production of manufactured goods. Given that manufacturing was an essential ingredient of the population and economic booms that occurred in Rust Belt cities during the nineteenth and early twentieth centuries (Beauregard 2006), deindustrialization left vast scores of urban workers in the Northeast and

Midwest unemployed. One response to the massive relocation and out-migration of manufacturing jobs was commensurate relocation and out-migration of people (population shrinkage).

Second, **suburbanization** involves the outward, horizontal expansion of built-up land relative to a central city. As new modes of transit enabled some urban workers to distance themselves from employment centers and high-density housing conditions in central cities, such workers responded by buying larger lots, and building larger homes, at the urban fringe. This conversion of open space into comparatively low density residential – and, later, commercial and other – land uses attracted critical masses of urban residents out of central cities in search of higher qualities of life (Ch. 5).

Third, natural **demographic change** operates through a place's birth rate, death rate, larger death events (e.g., natural disasters), and ageing and age structure. In addition to these natural tendencies, demographic change is also a function of migration – including the migration that results from deindustrialization and suburbanization. Migration is an obvious staple of population shrinkage. If the total number of out-migrants from a place is not matched by an equal or greater number of in-migrants, then population loss seems inevitable. Perhaps more subtly, though, beyond the absolute number of in- and out-migrants, the age structure of those migrants can have important feedback effects on urban change. For example, it is typical for a meaningful fraction of recent high school graduates to emigrate out of their home cities to attend college in other cities. Such out-migration shrinks a segment of the population that would otherwise make direct contributions to the birth rate and workforce of their home cities (Weber 2010). Therefore, it is not just the total number of people living in a city that matter for shrinkage but also the age structure of that population. Ageing affects urban shrinkage by removing retired people from the workforce, thereby lowering the overall income being produced per capita in an ageing city (Weber 2010). Additionally, an ageing population leads to increased dependence on younger populations. If a younger care-giver – say the adult son or daughter of an ageing resident – is living in a different city from the ageing person, then the older individual has an increased likelihood of emigrating out of their current city (and immigrating to the care-giver's city to receive care). In this case, like deindustrialization and suburbanization, ageing also results in migration.

Together, deindustrialization, suburbanization, and demographic change unquestionably reshaped the landscape of the Rust Belt (see Beauregard 2006). However, one issue with the notion that these three structural forces can be used to explain shrinkage and decline *in general* is that similar levels of each do not produce uniform outcomes. Presumably, much of the Rust Belt – especially places located within the highlighted area in Figure 4.1 – experienced deindustrialization, suburbanization, and demographic change with relatively comparable force. In reality, of the 19,068 census tracts that intersect the "Core Area of Shrinkage and Decline", only 24.2 percent (4,617) met the adopted threshold of annual population loss to be classified as *shrinking* (Fig. 2.2); and only 29.4 percent (1,356) of these shrinking tracts also endured *decline* [in the form of atypical increases in concentrated disadvantage] between 1970 and 2010 (Fig. 3.2). So, why did only a fraction of shrinking tracts in the "Core Area" also experience decline during the observed forty-year period of shrinkage?

One plausible answer to this question is that additional factors may determine the direction of urban change (Turchin 2003), that is, such forces (e.g., deindustrialization, suburbanization) cause crises in affected places through their effects on social and political institutions (Turchin 2007). To the extent that the mix of institutions varies widely from place to place, this observation implies that growth or shrinkage is context-dependent (Kearns and Forrest 2000). In other words, generalizations about shrinking places based solely on trends in "de-economization, de-population, and de-urbanization" (Hannemann 2004) disregard consequential information about the places' functional, spatial, cultural, and socio-political contexts.

Second, "context" is not an aggregate property of cities. Whereas a familiar approach to understanding urban shrinkage involves generating typologies out of municipal- or metropolitan-level empirical data (Schilling and Mallach 2012), political and social circumstances exhibit as much or more variation *within* cities as between them (Reckien and Martinez-Fernandez 2011). Nonetheless, city-level typologies tend to treat municipalities as "black boxes" with relatively homogeneous contents. But if, as argued above, the direction of urban change depends on how suburbanization, depopulation, and economic transitions interact with local contextual factors, then one should expect such outcomes to manifest differently *within a single city*. Because heterogeneity is a defining feature of cities (Laursen 2008), it follows that shrinkage and decline will not apply uniformly to the whole of any one locality. On the contrary, even so-called "shrinking cities" exhibit patterns of growth and/or stasis alongside their characteristic spaces of population loss (see Table 2.4).

Finally, "growth and decline feed off each other" (Beauregard 1993: 21). Recognizing this reality leads to the important point that local shrinkage can be the result of system-wide adjustments to global processes, and we need to continually track patterns, understand processes, and (re)define interventions (e.g., policy). Below, explanations of intra-urban change are discussed to show the evolution of thought.

Selected geographical theories of intra-urban change

Intra-urban theories of urban shrinkage and decline are generally categorized into one of three schools of thought – ecological, subcultural, or political economy (Table 4.1). Among the earliest attempts at theorizing intra-urban change are the Chicago School of Sociology's **ecological models** (Park and Burgess 1925). These models posit that cities are made up of individuals who compete with one another for urban space, the result being that households self-segregate "according to their ability to [pay for] different sites and situations" (Knox and Pinch 2010: 157). Notable works that are linked to the ecological school include Burgess' (1925) Concentric Ring model, Hoyt's (1939) Sector model, a class of Neighborhood Stages/Life Cycle models often connected to the U.S. Department of Housing and Urban Development (see Metzger 2000), and the Filtering/Vacancy Chain models proposed by Hoyt (1933) and used in other income-succession-based frameworks (Grigsby et al. 1987). While these models have noteworthy differences between them (Table 4.1), they and other ecological models tend to follow the same broad line of reasoning. Specifically, purposive household-level location decisions produce neighborhood units that are characterized by relatively homogeneous *within-neighborhood* social and physical fabrics, and relatively heterogeneous *between-neighborhood* social and physical fabrics.

On this backdrop, elements of the physical urban fabric depreciate over time. This natural tendency has two mutually reinforcing consequences. First, it reduces the level of satisfaction that existing occupants have in their current housing situations, thereby creating demand for newer, less-depreciated housing units. Since undeveloped land is often scarce in the urban core, new housing units are typically constructed at the urban fringe. At the same time, lower-status households are priced into formerly higher-status neighborhoods by the depreciating real estate values. Where these two forces operate in unison, the social fabric of an affected neighborhood begins to change on a downward socio-economic trajectory, which leads – gradually, over sufficiently long time horizons – to a qualitatively different neighborhood characterized by a downgraded physical fabric and accompanying reductions in social investments and institutions (Temkin and Rohe 1998). Gordon (2008: 8) refers to this dynamic process as the "iron law of urban decay":

> [r]ising incomes breed suburbanization. Suburbanization robs inner cities of their tax base. Inner city concentrations of poverty widen gaps between urban residents and substantive economic opportunities, and between suburban residents and urban concerns. And all of this encourages more flight, not only from the metropolitan core, but from decaying inner suburbs as well.

While ecological models offer plausible explanations for neighborhood change and patterns of decline, critics note that the school of thought focuses too narrowly on exogenous forces – i.e., it claims that because housing depreciates as it ages, all neighborhoods eventually decline (refer to the so-called "law" of urban decay quoted above). In contrast, some argue that this tendency can be overpowered by collective action on the part of neighborhood residents (Firey 1945). That is, signs of disinvestment in a neighborhood trigger balancing feedback effects such as residents coming together in unified fronts against ecological transformation. This perspective, which is associated with a class of **subcultural models** of urban change, therefore implicitly recognizes that complex interactions and non-linearities are defining features of urban systems. Accordingly, subculturalists view neighborhood change as an endogenous process, and not, as suggested by the ecological school, fully determined by exogenous variables.

To subculturalists, then, neighborhoods can decline, remain stable, or even improve depending on the strength of local social ties. Notwithstanding this proposition's real-world verifiability (Goodwin 1979), scholars note that the subcultural school is not well-suited to theory-building or generalization (Pitkin 2001). It does not explicate a mechanism responsible for creating the social environments that can resist decline; nor does it pay sufficient attention to higher levels of analysis (Weaver and Bagchi-Sen 2014). With respect to the latter, the school's tenets seemingly imply that strengthening social ties alone will arrest or reverse neighborhood decline (Kitchen and Williams 2009). This outcome is likely untenable in practice, for to some extent the capacity for collective action depends on exogenous and higher-level factors such as the presence of existing, effective institutions (Temkin and Rohe 1998). This issue is implicitly recognized by a third school of thought, political economy, which examines the inter-relationships between capital mobility, inequality, and the fate of neighborhoods (Pacione 2003).

Like urban ecological models, **political economy models** frequently portray neighborhood change in exogenous terms. Unlike the ecological school, though, decline is not natural and inevitable to a political economist. Instead, it is brought about by failure of the free market to produce equitable outcomes in, for instance, housing opportunities (Sassen 1990). The argument is that market economies produce geographically uneven patterns of development. Capital is distributed to locations where the gains to be made are greatest. Namely:

> [t]he differential use of space by capital in search of profit creates a mosaic of inequality at all geographic levels from global to local. Consequently, at any one time certain . . . regions, cities and *localities will be in the throes of decline as a result of the retreat of capital investment*, while *others will be experiencing the impact of capital inflows*. At the metropolitan scale the outcome of this uneven development process is . . . *sociospatial variations in life quality*.
>
> (Pacione 2003: 316; emphases added)

The above passage links growth, shrinkage, and decline in a dynamic relationship whereby all three outcomes occur simultaneously in the same regional and local contexts. From this vantage point, the political economy school argues that, whether fueled by unbalanced power relations in which growth machines control political institutions (Logan and Molotch 1987), restructuring in response to globalization (Sassen 1990), or some combination of these, market mechanisms invariably create socio-spatial residential segregation. The most disadvantaged neighborhoods in these configurations then experience self-reinforcing shrinkage and decline, as they lack the resources needed to collectively resist the types of negative change predicted by ecological models (Pacione 2003; Galster, Cutsinger, and Malega 2006).

As Table 4.1 suggests, the ecological, subcultural, and political economy schools of thought have had varying degrees of influence on urban theory and policy over time. Being the most recent, political economy has been a leading input to urban social theory since the 1980s (Pacione 2003).[4] Nonetheless, its relative recency and strong influence on contemporary theory do not automatically make it "the" framework with which to study patterns of coupled urban shrinkage and decline. Indeed, like the ecological school, political economy tends to treat neighborhoods as internally homogeneous – considerations of within-neighborhood diversity are barely

Table 4.1 Overview and examples of the three schools of thought on neighborhood change

Earliest Period of Influence	*School of Thought and Example Authors*	*Summary*
1920s	Ecological models: Concentric Rings model (Burgess 1925)	A city is composed of six concentric rings: central business district, industrial, slums, working class housing, higher class housing, commuter housing. Growth of the city leads to lower classes expanding into formerly higher class housing areas.
	Sector model (Hoyt 1939)	Growth occurs in sectors determined largely by transportation arteries, physical features, previous land use patterns, and amenities. Sectors will maintain similar socio-economic characteristics as they expand outward along these corridors.
	Life Cycle models (Babcock 1932; HOLC 1940)	Neighborhoods inevitably trend toward decline. The Life Cycle theory has had significant impact on urban renewal planning and mortgage lending policies in the U.S.
	Vacancy Chain/Filtering models (Hoyt 1933; Grigsby et al. 1987)	As higher-income residents leave for newer, better-quality housing, lower-income residents replace them as their vacated homes become affordable. This leaves the oldest, lowest-quality housing either abandoned, or serving as very low-quality housing for the lowest-income residents.
1940s	Subcultural Theory (Firey 1945; Kolb 1954)	Social cohesiveness and community integrity influence the trajectory of neighborhoods. Strong social ties and presence of positive social influencers can serve to protect neighborhoods from decline by maintaining social order, property investment, etc. If this influence is disrupted or diluted, a cycle of decline may commence.
1980s – present	Political Economy: Urban Growth Machine (Logan and Molotch 1987)	Cycles of growth and decline are driven by an elite class who view place as a commodity to be developed to maximize profit and gain for the elites, often at the expense of lower-income residents who are displaced.

Spatial Mismatch (Kain 1968)	There is a disconnect between where the urban poor live and where the jobs they are qualified for are being created. New jobs being created in the city are high-value, high-education jobs and current residents often do not have the skills for those jobs. The service and manufacturing jobs have moved to the suburbs, which may not be accessible to the inner-city residents, leaving them with limited employment options.
Urban Restructuring (e.g., Sassen 1990)	Deindustrialization, capital concentration in "global" cities, and changes in employment, have left cities with significant challenges, including strained public sector budgets, high unemployment, abandoned buildings, and residents with limited access to employment opportunities.

addressed (Pitkin 2001). This feature is essentially the inverse of the critique of the subcultural school – that it concerns itself too much with micro-level behaviors while overlooking higher-level political/economic constraints (Kitchen and Williams 2009). All told, then, the three schools of thought face similar general criticisms: they have narrow theoretical foci on either internal or external influences but not both and inflexible empirical foci on single geographic levels of analysis. Consequently, there have been calls for, and attempts to establish, balanced approaches that (1) are "flexible"; (2) "recognize forces from both within and outside of neighborhoods"; and (3) "analyze change at multiple geographic" levels (Pitkin 2001: 20–22). The next section covers one such attempt in an abridged manner.

A social capital model of intra-urban change

In an effort to explicitly link the three aforementioned schools of thought, Temkin and Rohe (1998) put forward a model of intra-urban change, which postulates that a neighborhood's trajectory hinges on its internal supply of *social capital*. The concept of social capital receives ample attention, and thus a multitude of definitions exist from interdisciplinary social scientists (see: Portes 1998). Drawing on context, it appears that Temkin and Rohe (1998) follow prominent social capital theorist Robert Putnam in taking it to mean the "features of social organization, such as trust, norms, and networks that can . . . facilitat[e] coordinated action" (Putnam, Leonardi, and Nanetti 1993: 167). More concisely, **social capital** describes the capacity of a group or neighborhood to act collectively (Turchin 2003). Hence, with respect to urban growth, shrinkage, and decline, social capital is a stock resource that can be drawn upon by neighborhoods that are facing pressure to undergo qualitative change. Neighborhoods with sufficiently high social capital are able to act cooperatively against decline, while those with deficient stocks follow the paths of decline predicted by the ecological models.

Like all other resources, social capital exhibits an uneven geographic distribution. According to the model advanced by Temkin and Rohe (1998), the positive social capital available in any given geographic neighborhood is a function of that neighborhood's (1) socio-cultural milieu and (2) institutional infrastructure. **Socio-cultural milieu** is a construct made up of identity, interaction, linkages, affect, and opportunities. Namely, a strong socio-cultural milieu exists in a given space when: (a) residents consistently identify with that space and attribute to it some symbolic meaning; (b) residents passively and actively interact on regular bases; (c) the space is connected to, but not necessarily an extension of, areas "outside" of the neighborhood; (d) identity is accompanied by affective ties to the identifiable space (refer to [a]); and (e) there are opportunities to engage in social activity (e.g., shopping, recreation, and worship). **Institutional infrastructure** refers to the existence and quality of "formal organizations" in the neighborhood. More precisely, it relates at once to the presence of [for instance] neighborhood associations *and* the political efficacy of those organizations. When a neighborhood is characterized by both a strong socio-cultural milieu and efficacious institutions, it is said to possess the requisite stock of social capital to defend against exogenous (higher level) or endogenous sources of change. Such neighborhoods either remain stable or gentrify over time, while neighborhoods without adequate social capital experience downward succession, and, presumably, contribute to citywide perceptions of urban shrinkage and decline.

Perhaps the most significant contribution of the Temkin and Rohe (1998) model for present purposes is its specification of a variable, *social capital*, on which the "drivers" of shrinkage and decline – suburbanization, deindustrialization, and demographic change – operate to influence the direction of urban change. In this vein, the model offers a potential explanation for why vastly different outcomes in aggregate population occur in cities that exhibit similar trends in these variables. In particular, social capital varies spatially and in magnitude between and within cities. No two cities will display identical (local) patterns of social capital, nor will they share the same (global) stocks of the resource. For that reason, it is necessary to dig deeper into the *intra-urban* processes that produce patterns of these events. The Temkin and Rohe (1998) model suggests that *social capital* is a pivotal part of these *underlying processes*.

Thinking operationally about social capital for large-area analysis

The literature on social capital includes differing perspectives (see Portes 1998). It is well outside the purview of this book to engage with these debates and trace the complex history of the concept. Hence, in this section we merely sketch out several of the more consensual points about *social capital*, with an eye toward identifying indicators of the concept for empirical research.

First, the social capital of a geographic community is largely intangible (Flora, Flora, and Gasteyer 2015). It cannot be observed and measured directly. For that reason, researchers have proposed a variety of survey instruments (Chazdon and Lott 2010), participatory research designs (Allen et al. 2012), experimental and ethnographic studies (Barr, Ensminger, and Johnson 2010), and proxy variables (Temkin and Rohe 1998; Rupasingha, Goetz, and Freshwater 2006) for quantifying components or dimensions of social capital at different geographic levels. Among these methods of operationalization, surveys, participatory research, and experimental and ethnographic studies tend to be costly, time-consuming, and often restricted to manageable (narrow) study areas and population subgroups. However, if well-designed, they can be extremely effective at uncovering crucial context-specific qualitative social and behavioral (inter)actions that otherwise remain hidden in larger scope quantitative investigations. In contrast, the use of quantitative proxy variables from existing secondary datasets is relatively inexpensive and can typically be accomplished for extensive and diverse study areas such as the United States (Rupasingha, Goetz, and Freshwater 2006). Nevertheless, as suggested above, this strategy is impersonal: it generally cannot reveal as complete a picture of specific social relations as compared to a more qualitative approach.

The previous paragraph implicates at least one key trade-off in empirical social capital research – between maximizing depth and context-sensitivity in a relatively qualitative investigation on one hand and maximizing scope and [perhaps for that reason] generalizability with a more quantitative investigation on the other.[5] Recall that the proxy variable strategy carries low operational costs. For many political decision-makers, planners, and students or other low- or un-funded researchers, this attribute is attractive. Indeed, cost effectiveness is partly responsible for our own decision to study social capital with secondary data later in this subsection. More importantly than cost though, we focus on the scope in order to continue analyzing patterns of shrinkage and decline across the United States (Fig. 4.1). In this way, we aim to provide an example of conducting a large-area analysis of [aspects of] social capital so that readers with interests in multijurisdictional study areas (e.g., metropolitan regions) can augment their toolboxes with some of the techniques employed below.

Components of social capital

The social capital model of neighborhood change postulates two constitutive components of *social capital*: (1) socio-cultural milieu and (2) institutional infrastructure. Socio-cultural milieu encapsulates the ideas of *collective norms* and *trust* (Temkin and Rohe 1998). **Collective norms** are “shared patterns of behavior, beliefs, and practices that are acquired via social learning” (Henrich and Henrich 2007: 133). Trust is the degree of belief that one or more other persons will take a valued action. Trust is exemplified by risk-taking, such as consigning one’s own resources to members of a group under the belief that he or she will be able to extract resources of equal or greater value from the group at a later point in time (Camerer 2003). A specific form of trust that is relevant to social capital is *generalized* trust. **Generalized trust** is an "abstract preparedness to trust [and engage in actions with]" anonymous others (Stolle 2001: 205). If such a predisposition to cooperate with strangers exists in a community, then the potential for working collectively (e.g., pooling resources) to solve neighborhood-level problems is enhanced.

Next, the institutional infrastructure component of the social capital model relates to the concept of *social networks*. A **social network** is a set of relations between [groups of] people (Kadushin 2012). Just like financial capital, networks have value. Networks give individuals and groups access to resources and knowledge that they themselves do not possess. For instance, neighbors might keep a watchful eye on one's property when one is out of town (Putnam 2007). This type of social support is a product of **bonding networks**, which are characterized by dense internal connections between persons who share something (e.g., spatial location or identity) in common. As another example, one might benefit from information, say about an upcoming job opening, which originated with a proverbial friend of a friend. This type of resource comes via a **bridging network**, in which persons or groups are connected to parts of the social system to which they would not be exposed but for their relations with a broker (i.e., the friend who has the friend).

Together, social networks, collective norms, and trust come together to form what is sometimes called "Putnam's three-legged stool" (Rogers and Jarema 2015: 16), named for the American political scientist whose oft-cited definition of social capital mentions these three elements. Notably, Putnam's research (Putnam, Leonardi, and Nanetti 1993; Putnam 2000) is largely credited with placing social capital on (inter)national policy agendas. While this work is not without critics (Portes 1998), research on the history of social capital suggests that the elements that make up Putnam's "three-legged stool" have been constant features in definitions of the concept since it first appeared in the literature (Rogers and Jarema 2015). Thus, networks, norms, and trust are assumed here to be accepted constituents of social capital.

In addition to the components that form Putnam's three-legged stool, two other concepts feature quite regularly in discussions of social capital in the context of intra-urban change. First, **social cohesion** describes a society that "hangs together", where "conflict between societal goals and groups . . . are largely absent or minimal" (Kearns and Forrest 2000: 996). In other words, a cohesive society is one in which most people contribute to the collective well-being. From this perspective, social cohesion can be viewed as a "bottom-up process founded upon local *social capital*" (Forrest and Kearns 2001: 2137; emphasis added). Like social capital, social cohesion is a subject of ongoing re-theorization, and we will not engage with the concept's voluminous body of scholarship in this book (see Stanley 2003). Rather, we note only the widespread agreement throughout the literature that *inequality strongly undermines cohesiveness*. Put differently, inequality enhances the propensity for inter-group conflict and disruptive behavior within a given geographic space (Larsen 2014).

Second, while *diversity* is critical to the long-term health, sustainability, and resilience of systems (Page 2011), in the short term, and during periods of transition, diversity "challenges social [cohesion] and inhibits social capital" (Putnam 2007: 138). In brief, the capacity for collective action in a geographic community is sometimes a positive function of community homogeneity. Like individuals tend to cooperate with, and build trust in, like individuals. This observation exposes a "negative side" of social capital (Portes 1998). Namely, it is possible for networks of individuals to share norms and trust one another but in ways that lead them to act collectively toward exclusionary or other socially harmful ends. For instance, strong *bonding networks* might coordinate to keep "outsiders" from penetrating their social circles, even when these *bridging* interactions might link members of the group up to profitable economic or cultural opportunities (Portes 1998). In the following section, bonding and bridging networks, collective norms, trust, social cohesion, and community homogeneity are used for investigating the importance of social capital in understanding patterns of coupled shrinkage and decline in the United States.

Social capital indicators in shrinking and shrinking-declining tracts

Drawing on existing literature, this subsection leverages data from the Brown University LTDB, the most recent (2010–2014) American Community Survey (ACS), and Esri Business Analyst 2014 (Box 2.1), to obtain proxy measures for the components of social capital identified above. With those proxy measures, we address a central question at issue in this chapter: what are some contextual variables that might take on significantly different values in shrinking census tracts that did not decline, relative to shrinking tracts that did decline over the course of our study period? Because the contextual variables that we examine hereinafter all relate to social capital, we necessarily consider the more specific question: does variation in social capital, as the Temkin and Rohe (1998) model of intra-urban change suggests, appear to be a key factor in producing patterns of coupled urban shrinkage and decline?

To begin, note that, with one exception, the secondary data we rely on for proxy variables from this point forward are not available in the Brown University LTDB. Instead, we obtain the bulk of our data from alternate datasets. Consequently, we are effectively testing observable implications of the social capital model of neighborhood change. Namely, if social capital played a substantive role in forestalling decline in some shrinking tracts but not others, then we should be able to detect a legacy of these patterns in *current* indicators of social capital. More precisely, current levels of social-capital-related variables should be higher in tracts that shrank but did not decline, relative to tracts that both shrank and declined.

With that caveat in mind, we collected data for various indicator variables to facilitate a two-part analysis. First, ample research suggests that housing tenure is significantly correlated with neighborhood-level outcomes (O'Brien 2012). In particular, home*owners* are more likely than renters to develop *collective norms* of neighborhood investment, political participation, and other forms of civic engagement (Fischel 2001). For that reason, the percentage of occupied units inhabited by owners is an established indicator of social capital (Glaeser, Laibson, and Sacerdote

2002; Rupasingha, Goetz, and Freshwater 2006). Data on *owner-occupied housing units* are available for all of the time periods (1970–2010) covered by the Brown University LTDB. Accordingly, this variable is the one indicator for which we investigate change over time, rather than static differences in current values. More explicitly, we compare the difference in the mean percentage of owner-occupied units between the two tract types in 1970 to the corresponding difference in 2010. We hypothesize that the gap in the mean values of the ownership variable widened between the two time periods. In other words, we expect that tracts that shrank but did not decline exhibited stable or growing ownership rates (a constant or increasing stock of one type of social capital); while tracts that both shrank and declined exhibited decreasing ownership (an erosion of one type of social capital).

A useful method for testing this hypothesis is difference-in-differences (DiD) analysis (Gerber and Green 2012). For any given quantity of interest, a DiD estimator subtracts the mean difference in that quantity between two groups before some event from the mean between-group difference after the event. Any observed pre-event difference is presumably the result of unobserved variation in group-level attributes. Assuming that these unobserved attributes do not vary with time, the "before" difference between groups should equal the "after" difference, plus any effect from the event.

Table 4.2 presents the results from a DiD analysis for the ownership variable from the Brown University LTDB. The mean percentage of owner-occupants in tracts that shrank but did not decline was 54.6 percent in 1970. The comparable value for tracts that both shrank and declined was 51.5 percent. Hence, in 1970 the mean percentage of owner-occupants was 3.1 percentage points higher in the former type of tract relative to the latter type. If all shrinking tracts experienced relevant forces of intra-urban change (e.g., deindustrialization, suburbanization, and demographic change) similarly during the study period, then in 2010 we should also expect a difference in the mean percentage of the ownership variable of about 3.1 percentage points. In fact, while mean percentage of owner-occupants in the former tract type remained roughly the same (54.6 percent in 1970 and 54.8 percent in 2010), in the latter shrinking-declining type of tract, mean percentage of homeowners dropped from 51.5 percent in 1970 to 42.4 percent in 2010. Thus, the gap in mean percentage of owner-occupants between the two tract types widened

Table 4.2 Change in housing tenure, 1970–2010

Tract Type	*Owner Occupied, 1970 (Mean % of Occupied Units)*	*Owner Occupied, 2010 (Mean % of Occupied Units)*	*n*
Shrank Only	54.6	54.8	5,346
Shrank and Declined	51.5	42.4	1,849
Difference:	3.1	12.4	7,195
Difference-in-Differences:		9.4***†	

***$p << 0.001$ †pseudo p-value obtained via randomization inference (9999 permutations)

considerably, from 3.1 percentage points to 12.4 percentage points between 1970 and 2010.

The difference in these differences (9.4 percentage points) is our DiD estimate, which is highly statistically significant (pseudo $p << 0.001$). In other words, relative to tracts that shrank and did not decline, tracts that experienced coupled shrinkage and decline witnessed a substantial drop in homeownership rates that cannot be explained by chance alone. If homeownership is a useful indicator of social capital (Glaeser, Laibson, and Sacerdote 2002), then this result implies that social capital may have decreased in these shrinking-declining tracts as well. Thus, the tracts that are able to keep their homeownership rates steady during population shrinkage may be more resistant to decline via social capital. On the other hand, tracts in which population shrinkage is correlated with a reduction in homeownership – possibly because out-migrants were homeowners who previously contributed positively to neighborhood social capital – may be more likely to experience decline.

Next, given the results from Table 4.2, it follows that tracts that shrank but did not decline may exhibit greater *current* levels of social capital relative to shrinking-declining tracts. That is, the DiD analysis implies that homeownership, and by association social capital, has a legacy. Where homeownership is maintained at sufficiently high levels, places appear to be less likely to experience decline. Thus, tracts that resisted decline over our forty-year study period may have inherited (i.e., and may currently possess) relatively high stocks of social capital, and conversely for tracts that declined. On that backdrop, we obtained current data on six indicator variables to (imperfectly) measure the five components of social capital discussed above.

First, following Putnam (1993) and Rupasingha, Goetz, and Freshwater (2006), certain types of incorporated organizations convey information about the institutions and *networks* that are present in geographic communities. In particular, the number of "Religious, Grantmaking, Civic, Professional, and Similar Organizations" in a given space is a rough indicator of the formal collective-minded bonding networks that exist in that location. Within the North American Industry Classification System (NAICS), all such (hereinafter "civic") organizations begin with the three-digit code "813" (U.S. Census Bureau n.d.). Accordingly, we extracted all civic organizations with the NAICS prefix "813" from Esri Business Analyst 2014 (Box 2.1), a dataset that provides comprehensive information on the locations of incorporated entities in the United States. The civic organizations were then joined and aggregated to our geospatial census tract dataset (see Ch. 2). To account for differences in both population and area, we then created a measure of *civic organizations per 1,000 persons (per square mile)* to serve as a proxy for bonding network connections within census tracts.

Second, again using the Esri Business Analyst dataset and by way of similar extraction and computational techniques, we derived a measure of *government offices per 1,000 persons (per square mile)* for each census tract.[6] One resource that is regularly accessed through *bridging networks* is political capital (Flora, Flora, and Gasteyer 2015). The greater the presence and accessibility of government offices within a given census tract neighborhood, the more potential opportunities there are for persons living in that neighborhood to interact with public decision-makers who might not otherwise be part of their social networks. Of course, this logic assumes that all persons are equally likely to get through the doors of government offices, which is certainly not the case. Thus, the true degree to which more government offices translates into greater accessibility to government resources cannot be evaluated with our large-scale secondary data sources. For that reason, we are left with an admittedly weak surrogate measure of bridging networks but one that allows us to proceed with our investigation.

Third, we have already noted that homeownership is an established indicator of social capital (Glaeser, Laibson, and Sacerdote 2002), in that it suggests the existence of at least some *collective norms* (Fischel 2001). Taking this argument a step further, given that collective norms are a product of "social learning" (Henrich and Henrich 2007), one should expect norms to be stronger the longer individuals interact in environments where social learning takes place. In that context, for each census tract we measure the percentage of occupied housing units that are inhabited by owners who have lived in their current units for a minimum of twenty-five years (i.e., they have been at their residences since at least 1989). This measure of *long -term owners* comes directly from the current U.S. Census ACS, which reports data on housing tenure by the year households moved into their residences.

Fourth, in a study of cultural identity in the American Appalachian region, Weaver and Holtkamp (2016) proposed that expenditures on used goods contain at least some information on *generalized trust* in society.[7] The sellers (perhaps including intermediate sellers) of used goods always have more information about the quality of those goods than prospective buyers. Hence, the act of purchasing a used good is, albeit very weakly (Weaver and Holtkamp 2016), a sign that a used-good buyer *trusts* that the used-good seller did not knowingly put a damaged or dysfunctional good up for sale on the market. For all geographic units in its dataset, Esri Business Analyst provides data on the local "retail potential" of a given economic sector, where local retail potential is tantamount to household expenditures in that sector in the overall economy (Esri 2014). For each census tract in our dataset, we therefore extracted the total volume of these used-good expenditures from the Esri dataset and divided by the tract's total population to obtain a measure of *used-good expenditures per capita*.

The first four variables described above are all hypothesized to relate positively to social capital. In other words, the higher the value of the variable, the higher the (assumed) stock of social capital. In contrast, the final two variables included in this analysis are hypothesized to relate negatively to social capital. The fifth variable, *income inequality*, is measured with the popular Gini coefficient (Todaro and Smith 2014). The Gini coefficient of income inequality ranges from 0 to 100 (when multiplied by 100, which we have done), where higher values indicate less-equitable distributions of income. The Gini coefficient is available directly from the U.S. Census ACS, and it is a widely used measure of inequality in both the United States and the international community.[8] Insofar as inequality is known to undermine *social cohesion*, a higher Gini coefficient is suggestive of lower social capital.

Finally, recall that in the short term, and during periods of transition such as shrinkage and decline, ethnic diversity is correlated with lower social capital. By extension, *community homogeneity* is linked to higher social capital in these cases (Putnam 2007). For every geographic unit in the dataset, Esri Business Analyst reports an interaction-based *ethnic diversity index*, which measures the probability that two randomly selected persons from a given unit of analysis (for us, a census tract) are members of different racial or ethnic groups. The index ranges from 0 to 100, where higher values indicate more diversity or less homogeneity (Reese-Cassal 2014).

Similar to several other variables that have been examined to this point in the book, the six proxy measures described above are characterized by skewed distributions. For that reason, we evaluate differences in the variables using Wilcoxon signed-rank tests, which loosely compare the medians of two independent samples. Table 4.3 presents the results of these tests for each of our six indicators. As expected, the median values of the first four indicators – for bonding connections, bridging connections, collective norms, and generalized trust, respectively – are all statistically significantly higher in tracts that experienced population shrinkage *without* decline, compared to tracts that endured both shrinkage *and* decline. Likewise, the median values of the final two indicators – income inequality and ethnic diversity – are statistically significantly lower in "Shrink Only" tracts relative to the "Shrink and Decline" tracts.

Because social capital is an intangible asset that cannot be directly observed and measured (Flora, Flora, and Gasteyer 2015), this subsection employed multiple quantitative indicators in what was effectively a weights-of-evidence type of investigation. More plainly, we evaluated the same underlying hypothesis with seven different variables and two different analyses (Tables 4.2 and 4.3), with the hope that the separate tests would collectively implicate a link between social capital indicators

Table 4.3 Wilcoxon signed-rank tests for equality of medians in independent samples

Variable	*Shrink Only*	*Shrink and Decline*	*Difference*	*Indicator of:*
Civic Organizations (per 1,000 persons per square mile)	2.4	1.8	0.6***	Bonding Connection
Government Offices (per 1,000 persons per square mile)	0.3	0.2	0.1***	Bridging Connection
Long-Term Owners (% of occupied units)	19.1	16.4	2.7***	Collective Norms
Used-Good Expenditures Per Capita (US$)	25.0	14.9	10.1***	Generalized Trust
Income Inequality Index	43.2	45.4	–2.2***	Social Cohesion (–)
Ethnic Diversity Index	38.9	48.7	–9.8***	Homogeneity (–)
n[a]	5,253	1,839		

[a] Excludes cases with missing data ***$p << 0.001$

and neighborhood outcomes. Indeed, such a link seems to manifest in our results. Compared to their declining counterparts in our study, shrinking census tracts that capably withstood pressures to undergo severe qualitative decline now seem to have:

- stable homeownership rates;
- a relatively high number of civic organizations;
- a relatively high number of government offices;
- a greater share of long-term stakeholders;
- a greater predisposition to trust others in transactions characterized by asymmetric information;
- slightly lower income inequality; and
- greater internal community homogeneity.

If only one or a few of these hypotheses panned out, then it would be difficult to claim a connection between social capital and tract-level outcomes vis-à-vis shrinkage and decline. That all results are significant and in the expected directions, however, makes a persuasive circumstantial case that to better understand patterns of shrinkage and decline, researchers, planners, community activists, political decision-makers, and other stakeholders would do well to view those patterns in concert with local variations in *social capital*. Engagements with deindustrialization, suburbanization, and demographic change are unlikely to be sufficient on their own for explaining the variability in intra-urban patterns of growth, stability, shrinkage, and decline. Rather, as the model proffered by Temkin and Rohe (1998) posits, and as the results from Tables 4.2–4.3 imply, social capital appears to be an important aspect of intra-urban change.

That being said, the findings in this section should be considered as starting points and not as conclusive proof that high stocks of social capital, as measured through various stock indicators, necessarily fend off downward spiral. While we followed established practices for operationalizing and analyzing various components of social capital (Putnam, Leonardi, and Nanetti 1993; Putnam 2000; Rupasingha, Goetz, and Freshwater 2006), discovering that census tracts that declined may possess ostensibly lower social capital than tracts that did not decline "is not particularly helpful unless one understands *why* some places are able to develop stronger [social capital] than others" (Brown and Schafft 2011: 228). In other words, discussions of social capital *stocks* are largely missing a discussion of the factors or conditions that facilitate stock accumulation.

Statistical associations between social capital stocks and outcomes can be very useful for placing social capital on political agendas, as Putnam's work seems to have done at a national level (Portes 1998). Nevertheless, a crucial next step at the local and intra-urban levels is to identify ways of building efficacious and inclusive social networks, facilitating the creation of shared prosocial norms, and creating a sense of trust and collective mindedness within communities. Among other tactics, recent research suggests that participatory planning and community strategic visioning are useful tools for establishing these components of social capital in intra-urban settings (Walzer and Hamm 2012). More generally, efforts to enhance the degree to which diverse stakeholders communicate with one another (Ostrom 2003) and are able to participate meaningfully in collective decision-making, monitoring, and enforcement activities (Ostrom 1990) are effective means for building trust, norms, and efficacious networks in neighborhoods. Several of these ideas will be revisited in Chapter 7.

Concluding remarks

This chapter shows that urban change is context-dependent. There is as much or more intra-place variation as there is inter-place variation. This means that even "shrinking cities" contain spaces of growth and stability within their boundaries. Thus, typology-based approaches to theorizing about urban shrinkage and decline at the city level can result in the loss of critical information about actually existing shrinkage and decline. To bring these ideas to bear on studies of shrinkage and decline, this chapter surveyed selected theories of intra-urban change. The literature review draws attention to the variation in *social capital* as a crucial part of the explanation of shrinkage and decline – that is, the local context is also contributing to the production of *patterns and trends* in urban shrinkage and decline. To assess the validity of this statement, the chapter presented results from data analyses that examined differences in multiple indicators of social capital, between shrinking census tracts that did and did not undergo decline during the analyzed period. The results uniformly point to the conclusion that variation in stocks of social capital is correlated with variation in neighborhood outcomes. Shrinking census tracts that resist severe decline seemingly possess higher stocks of social capital relative to tracts that shrink and decline in unison. Several policy implications of these findings are explored in Chapter 7. In the nearer term (Ch. 5), we turn attention to expanding our understanding of the conditions that produce patterns of shrinkage and decline in the city (*urban*) to include explanations for how (or if) these patterns have begun, or are beginning, to form in *suburban* settings.

Notes

1 See: www.latimes.com/nation/la-na-west-baltimore-profile-20150428-story.html
2 See: http://www.msnbc.com/interactives/geography-of-poverty/ne.html
3 See: http://www.nytimes.com/2013/07/19/us/detroit-files-for-bankruptcy.html?pagewanted=all&_r=0
4 However, its critiques of uneven development and neoliberal policy instruments are not particularly well-reflected in Western urban policy (for examples, see the volume edited by Brenner and Theodore 2002).
5 NB: where feasible, a mix of these strategies is certainly desirable (e.g., Glaeser, Laibson, and Sacerdote 2002).
6 Within the Standard Industrial Classification (SIC) system, all government offices begin with the two-digit code "91". All such entities were extracted from Esri Business Analyst 2014 to create this variable.
7 NB: Weaver and Holtkamp (2016) found that used-goods sales are not positively correlated with low income, which suggests that the former is not simply an expression of, or surrogate for, the latter.
8 See: http://hdr.undp.org/en/content/income-gini-coefficient

References

Allen, Ryan, Scott Chazdon, Barbara Radke, and Tobias Spanier. 2012. "Ready to Vision?: Evidence from Social Capital Assessments in Four Minnesota Towns." In *Community Visioning Programs: Processes and Outcomes*, edited by Normal Walzer and Gisele Hamm, 53–73. New York: Routledge.

Barr, Abigail, Jean Ensminger, and Jeffrey C. Johnson. 2010. "Social Networks and Trust in Cross-Cultural Economic Experiments." In *Whom Can We Trust? How Groups, Networks, and Institutions Make Trust Possible*, edited by Karen S. Cook, Margaret Levi and Russell Hardin, 65–90. New York: Russell Sage Foundation Press.

Beauregard, Robert A. 1993. "Representing Urban Decline Postwar Cities as Narrative Objects." *Urban Affairs Review* 29 (2):187–202.

Beauregard, Robert A. 2006. *When America Became Suburban*. Minneapolis: University of Minnesota Press.

Beauregard, Robert A. 2009. "Urban Population Loss in Historical Perspective: United States, 1820–2000." *Environment and Planning A* 41 (3):514–528.

Bluestone, Barry, Mary Huff Stevenson, and Russell Williams. 2008. *The Urban Experience: Economics, Society, and Public Policy*. Oxford: Oxford University Press.

Brenner, Neil, and Nik Theodore, eds. 2002. *Spaces of Neoliberalism: Urban Restructuring in Western Europe and North America*. Maiden, MA: Blackwell Publishers Ltd.

Brown, David L., and Kai A. Schafft. 2011. *Rural People and Communities in the 21st Century: Resilience and Transformation*. Cambridge: Polity Press.

Burgess, Ernest. 1925. "The Growth of the City: An Introduction to a Research Project." In *The Trend of Population*, 85–97. Chicago: Publications of the American Sociological Society.

Camerer, Colin. 2003. *Behavioral Game Theory: Experiments in Strategic Interaction*. Princeton: Princeton University Press.

Chazdon, Scott A., and Stephanie Lott. 2010. "Ready for Engagement: Using Key Informant Interviews to Measure Community Social Capacity." *Community Development* 41 (2):156–175.

Esri. 2014. "Business Analyst." Esri. http://doc.arcgis.com/en/bao/.

Firey, Walter. 1945. "Sentiment and Symbolism as Ecological Variables." *American Sociological Review* 10 (2):140–148.

Fischel, William A. 2001. *The Homevoter Hypothesis: How Home Values Influence Local Government Taxation, School Finance, and Land-Use Policies*. Cambridge: Harvard University Press.

Flora, Cornelia Butler, Jan L. Flora, and Stephen Gasteyer. 2015. *Rural Communities: Legacy + Change*. 5th ed. Boulder: Westview Press.

Forrest, Ray, and Ade Kearns. 2001. "Social Cohesion, Social Capital and the Neighbourhood." *Urban Studies* 38 (12):2125–2143.

Galster, George C, Jackie M Cutsinger, and Ron Malega. 2006. "The Social Costs of Concentrated Poverty: Externalities to Neighboring Households and Property Owners and the Dynamics of Decline." Revisiting Rental Housing: A National Policy Summit, Joint Center for Housing Studies, Harvard University, Cambridge, MA, November 14–15.

Gerber, Alan S., and Donald P. Green. 2012. *Field Experiments: Design, Analysis, and Interpretation*. New York: WW Norton.

Glaeser, Edward L., David Laibson, and Bruce Sacerdote. 2002. "An Economic Approach to Social Capital*." *The Economic Journal* 112 (483):437–458.

Goodwin, Carole. 1979. *The Oak Park Strategy: Community Control of Racial Change*. Chicago: University of Chicago Press.

Gordon, Colin. 2008. "Patchwork Metropolis: Fragmented Governance and Urban Decline in Greater St. Louis." *Saint Louis University Public Law Review* 34 (1):21.

Grigsby, William, Morton Baratz, George Galster, and Duncan Maclennan. 1987. "The Dynamic of Neighborhood Change and Decline." *Progress in Planning* 28:1.

Großmann, Katrin, Marco Bontje, Annegret Haase, and Vlad Mykhnenko. 2013. "Shrinking Cities: Notes for the Further Research Agenda." *Cities* 35:221–225.

Hannemann, Christine. 2004. *Marginalisierte Städte: Probleme, Differenzierungen Und Chancen Ostdeutscher Kleinstädte Im Schrumpfungsprozess*. Berlin: BWV, Berliner Wissenschafts-Verlag.

Henrich, Joseph, and Natalie Henrich. 2007. *Why Humans Cooperate: A Cultural and Evolutionary Explanation*. Oxford: Oxford University Press.

Hospers, Gert-Jan. 2014. "Urban Shrinkage in the EU." In *Shrinking Cities: A Global Perspective*, edited by Harry W. Richardson and Chang Woon Nam, 47–58. London: Routledge.

Hoyt, Homer. 1933. *One Hundred Years of Land Values in Chicago: The Relationship of the Growth of Chicago to Its Land Values, 1830–1933*. Chicago: The University of Chicago Press.

Hoyt, Homer. 1939. *The Structure and Growth of Residential Neighborhoods in American Cities*. Washington, DC: Federal Housing Administration.

Kadushin, Charles. 2012. *Understanding Social Networks: Theories, Concepts, and Findings*. Oxford: Oxford University Press.

Kearns, Ade, and Ray Forrest. 2000. "Social Cohesion and Multilevel Urban Governance." *Urban Studies* 37 (5/6):995.

Kitchen, Peter, and Allison Williams. 2009. "Measuring Neighborhood Social Change in Saskatoon, Canada: A Geographic Analysis." *Urban Geography* 30 (3):261–288.

Knox, Paul, and Steven Pinch. 2010. *Urban Social Geography: An Introduction*. 6th ed. New York: Prentice Hall.

Larsen, Christian Albrekt. 2014. *Social Cohesion: Definition, Measurement and Developments*. Aalborg University, Denmark: Centre for Comparative Welfare Studies.

Laursen, Lea Louise Holst. 2008. *Shrinking Cities or Urban Transformation*. Aalborg: Aalborg University Press.

Lee, Jay, and Davis W. S. Wong 2001. *Statistical Analysis with ArcView GIS*. New York: John Wiley & Sons, Inc.

Logan, John, and Harvey Molotch. 1987. *Urban Fortunes: The Political Economy of Place*. Berkeley: University of California Press.

Metzger, John T. 2000. "Planned Abandonment: The Neighborhood Life-Cycle Theory and National Urban Policy." *Housing Policy Debate* 11 (1):7–40.

O'Brien, Daniel Tumminelli. 2012. "Managing the Urban Commons." *Human Nature* 23 (4):467–489.

Ostrom, Elinor. 1990. *Governing the Commons: The Evolution of Institutions for Collective Action*. Cambridge: Cambridge University Press.

Ostrom, Elinor. 2003. "Toward a Behavioral Theory Linking Trust, Reciprocity, and Reputation." In *Trust and Reciprocity: Interdisciplinary Lessons from Experimental Research*, edited by Elinor Ostrom and James Walker, 19–79. New York: The Russell Sage Foundation.

Pacione, Michael. 2003. "Quality-of-Life Research in Urban Geography." *Urban Geography* 24 (4):314–339.

Page, Scott E. 2011. *Diversity and Complexity*. Princeton: Princeton University Press.

Park, Robert E., and Ernest W. Burgess, eds. 1925. *The City*. Chicago: The University of Chicago Press.

Pitkin, Bill. 2001. *Theories of Neighborhood Change: Implications for Community Development Policy and Practice*. Los Angeles: UCLA Advanced Policy Institute.

Portes, Alejandro. 1998. "Social Capital: Its Origins and Applications in Modern Society." *Annual Review of Sociology* 24:1–25.

Putnam, Robert D. 1993. "The Prosperous Community." *The American Prospect* 13:35–42.

Putnam, Robert D. 2000. *Bowling Alone: The Collapse and Revival of American Community*. New York: Simon & Schuster.

Putnam, Robert D. 2007. "E Pluribus Unum: Diversity and Community in the Twenty-First Century the 2006 Johan Skytte Prize Lecture." *Scandinavian Political Studies* 30 (2):137–174.

Putnam, Robert D., Robert Leonardi, and Raffaella Y. Nanetti. 1993. *Making Democracy Work: Civic Traditions in Modern Italy*. Princeton: Princeton University Press.

Reckien, Diana, and Cristina Martinez-Fernandez. 2011. "Why Do Cities Shrink?" *European Planning Studies* 19 (8):1375–1397.

Reese-Cassal, Kyle. 2014. *2014/2019 Esri Diversity Index*. Redlands: Esri.

Rogers, Shannon H., and Patricia M. Jarema. 2015. "A Brief History of Social Capital Research." In *Social Capital at the Community Level: An Applied Interdisciplinary Perspective*, edited by John M. Halstead and Steven C. Deller, 14. New York: Routledge.

Rupasingha, Anil, Stephan J. Goetz, and David Freshwater. 2006. "The Production of Social Capital in US Counties." *The Journal of Socio-Economics* 35 (1):83–101.

Sassen, Saskia. 1990. "Economic Restructuring and the American City." *Annual Review of Sociology* 16:465–490.

Schilling, Joseph M., and Alan Mallach. 2012. *Cities in Transition: A Guide for Practicing Planners*. Vol. 568, *Planning Advisory Service*. Chicago: American Planning Association.

Stanley, Dick. 2003. "What Do We Know about Social Cohesion: The Research Perspective of the Federal Government's Social Cohesion Research Network." *The Canadian Journal of Sociology* 28 (1):5–17.

Stolle, Dietlind. 2001. "Getting to Trust: An Analysis of the Importance of Institutions, Families, Personal Experiences and Group Membership." In *Politics in Everyday Life: Social Capital and Participation*, edited by Paul Dekker and Eric M. Uslaner, 118–133. New York: Routledge.

Temkin, Kenneth, and William M. Rohe. 1998. "Social Capital and Neighborhood Stability: An Empirical Investigation." *Housing Policy Debate* 9 (1):61–88.

Todaro, Michael P., and Stephen C. Smith. 2014. *Economic Development, The Pearson Series in Economics*. Philadelphia: Trans-Atlantic Publications.

Turchin, Peter. 2003. *Complex Population Dynamics: A Theoretical/Empirical Synthesis*. Vol. 35, *Monographs in Population Biology*. Princeton: Princeton University Press.

Turchin, Peter. 2007. *War and Peace and War: The Rise and Fall of Empires*. New York: Plume.

Walzer, Norman, and Gisele F. Hamm. 2012. *Community Visioning Programs: Processes and Outcomes*. New York: Routledge.

Weaver, Russell, and Chris Holtkamp. 2016. "Determinants of Appalachian Identity: Using Vernacular Traces to Study Cultural Geographies of an American Region." *Annals of the American Association of Geographers* 106 (1):203–221.

Weaver, Russell C., and Sharmistha Bagchi-Sen. 2014. "Evolutionary Analysis of Neighborhood Decline Using Multilevel Selection Theory." *Annals of the Association of American Geographers* 104 (4):765–783.

Weber, Lars. 2010. *Demographic Change and Economic Growth: Simulations on Growth Models, Contributions to Economics*. Berlin: Springer-Verlag.

5 Shrinkage and decline beyond the central city

In this chapter we explore several of the theoretical economic foundations of two main drivers of shrinkage and decline: suburbanization and deindustrialization. By engaging with the ways in which economic incentives influence (1) household decisions to migrate from central cities to suburbs and (2) manufacturing firm decisions to relocate plants from older urban cores to more profitable spaces, we begin to appreciate just why suburbanization and deindustrialization are repeatedly cited as the main "causes" of population shrinkage (Großmann et al. 2013). Aligning with the arguments and findings from Chapter 4, however, to view shrinkage and decline as deterministic outcomes of suburbanization and deindustrialization masks important variation within the larger metropolitan system. For example, *suburbanization* is often said to cause *urban* shrinkage and decline through a dynamic process in which households (and their wealth) are simultaneously pushed and pulled beyond central city borders in search of higher qualities of life. This well-told story tends to imply that shrinkage and decline are predominantly urban problems: cities hollow out and undergo negative qualitative change, while their surrounding suburbs grow and prosper.[1] Maintaining this sort of city-suburb dichotomy diverts attention from important patterns of *actually existing* shrinkage and decline, which can and do occur beyond central city borders in the United States. We once again draw on longitudinal census tract-level data to demonstrate these patterns and present findings that suggest that the phenomena of shrinkage, decline, and coupled shrinkage and decline could be in the beginning stages of a longer-term shift away from central cities toward the inner suburbs and beyond.

To provide an example of how these shifting patterns might manifest geographically, we present a brief illustrative case study of the Greater St. Louis metropolitan region. The case study maps forty-year changes (1970–2010) in the spatial distribution of census tracts that exhibit high degrees of *concentrated disadvantage* (Ch. 3) in the study area. This exercise reveals that relatively recent patterns of *decline* – defined as decline that occurred over the past decade – have spilled into suburban communities immediately surrounding the central city (so-called "first ring" suburbs). Thus, the case study demonstrates that suburbs are not exempt from the "urban" issues of shrinkage and decline that are said to be caused by suburbanization. On that foundation, the end of the chapter replicates our analysis of *social capital* variables from Table 4.3 for the full set of *non-core-city* census tracts from the Brown University LTDB that experienced recent (over the period 2000–2010) population shrinkage. Given our prior findings about the association of social capital with urban shrinkage and decline (Ch. 4), we argue that this important contextual variable is much needed to understand the broader patterns of shrinkage and decline.

A basic model of urban land expansion

To understand how places beyond central cities experience shrinkage and decline, it is helpful to first consider how those places initially grew. A simple but popular model that grapples with this question is the **bid-rent model** of urban economics. The bid-rent model was proposed in the 1960s by William Alonso to explain patterns of metropolitan land values and the distribution of land across different classes of end users (e.g., businesses, high-income households, low-income households, etc.). Although Alonso addressed several possible patterns of urban land use in his work, his model is probably best known for its application to **monocentric cities**. Monocentric cities are a classic pattern of urban development in which there is a high-density city center characterized by high land use intensity; and where the city center is surrounded by roughly concentric circles of decreasing land use intensity (e.g., culminating with a zone of large lot, single-family residences). As it turns out, many older American "shrinking" cities exhibit development patterns that approximate these abstract theoretical expectations (Jessen 2012). Accordingly, the *bid-rent model*, as it is applied to monocentric cities, is an accessible starting point for thinking about (sub)urban settlement and development patterns.

The bid-rent model is grounded in an apparent trade-off between one's accessibility to the urban core (city center) and the quantity of land owned. All else being equal, the model assumes that the most desired location in a monocentric city is the core central business district (CBD). Prior to the era of **mass automobility** (i.e. relatively ubiquitous automobile ownership; discussion to follow), a city's major employers clustered in the CBD. For producers of goods, location in the CBD minimized transportation costs. Manufacturers could produce and sell goods in the same (central) location, which meant they did not have to transport the bulk of their products to faraway marketplaces. Because cost minimization positively affects a company's profits, companies naturally demanded land in or near the CBD for these reasons. Moreover, firms typically possess more capital assets than households, so they could generally out *bid* prospective residential users for CBD lands during the early stages of a city's development. In this way, CBDs became hubs of commerce and industry.

Meanwhile, residential land users depended on the urban labor market for jobs and wages. As firms spatially concentrated in city centers, the labor markets of CBDs became larger and more robust. Job growth acted as a pull factor that drew rural residents into cities. Much as producers sought to minimize the costs of transporting their manufactured goods to a marketplace, participants in urban labor markets sought to minimize commuting costs between their workplace and home. In the early stages of city development – prior to mass transit and the rise of the automobile – most urban households demanded land as close to the CBD as possible. With both firms and households creating demand for CBD land in large quantities, cities underwent rapid expansion in two dimensions. First, CBDs expanded vertically, as it became profitable to use land near the city center as intensely as possible. Second, the relative scarcity of CBD land supply compared to the large demand for CBD land led to a horizontal (i.e., outward) expansion into the urban fringe. Whereas commercial and industrial firms outbid households for the most profitable CBD lands, households outbid other claimants (e.g., agricultural users) for the lands immediately surrounding the CBD.

Over time, as new transportation technologies such as streetcars and automobiles (see below) became available, many residential urban land users saw opportunities to distance themselves from the high-intensity land use and congestion (i.e., push factors) that existed in the CBD. These households traveled beyond the already built-up areas of cities to acquire undeveloped land at the fringe. While relocating to the urban fringe increased transportation costs between work and home, the amount of land that could be owned (and the separation from ostensible urban "nuisances") offset the increased costs for many relatively high-income households. At the same time, however, relatively low-income households, for whom the increased transportation costs of moving to the urban fringe were uneconomic, remained close to the CBD.

This sorting process is illustrated in Figure 5.1. The graph represents the *bid* that users are willing to pay for *land rent* at various distances from the CBD. The line labeled "CD" is the **bid-rent curve** for typical low-income households. Under

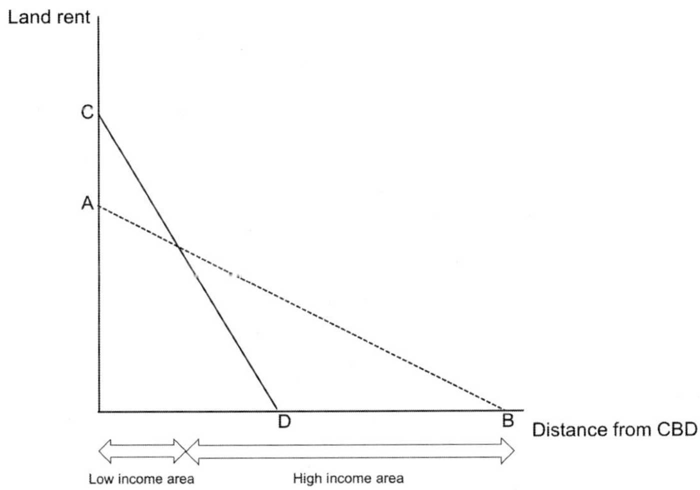

Figure 5.1 Hypothetical bid-rent curves for low-income (line CD) and high-income (line AB) households (adapted from Bluestone, Huff Stevenson, and Williams 2008)

the assumption that low-income households are unable to absorb the high transportation costs associated with living far from where they work (i.e., the CBD), low-income households tend to "outbid" (or out-demand) high-income households for residential land in, or surrounding, the CBD. On the other hand, high-income households are typically willing and able to trade-off higher transportation costs for more land and newer housing at the urban fringe. Such households therefore outbid lower-income residential users for land in these territories. The **bid-rent curve** for higher income households is labeled "AB" in Figure 5.1. The general pattern that emerges from these two bid-rent curves promotes residential income-based segregation: high-income households convert undeveloped land at the urban fringe into residential use; and, in doing so, they separate themselves from the low-income households who remain in close proximity to the commercial and industrial employment opportunities in the CBD.

In formulating the bid-rent model, Alonso called attention to what appears to be a **residential paradox.** The paradox is that, within the model, low-income households occupy some of the highest-valued urban land in or near the CBD, and high-income households occupy the cheaper land at the urban fringe. To resolve this paradox, Alonso made two key observations. First, transportation costs represent a large and increasingly burdensome percentage of low-income households' budgets the farther they live from their center city workplaces. More explicitly, the decrease in the price of land associated with distance from the city center does not compensate for the rising transportation costs incurred by low-income households in the bid-rent model. Second, residential land uses in or near the CBD occur at much higher densities relative to residential uses at the urban fringe. These high-density residential living arrangements – and the older and smaller dwelling units to which they are predominantly connected – are less frequently demanded by high-income households. The trade-off is therefore between low transportation costs and larger quantities of land and housing. High-income households in the model have the capacity to take on added transportation costs to consume more land and housing farther from the CBD, while low-income households do not.

On this backdrop, one outcome of residential land transactions in the bid-rent model is the creation of *suburban* communities. That said, whereas early phases of *suburbanization* outwardly followed the trajectories described above, changes in transportation technologies and the geographies of labor markets eventually made it possible for more diverse mixes of households to relocate to suburban communities. Thus, the common perception that suburbs have always been more or less homogenous and prosperous is a vast overgeneralization, and it hides the complex challenges that many suburban (and rural) communities are currently facing with respect to shrinkage and decline. The next section briefly unpacks key phases in the history of urban expansion in the United States, starting with the processes described by the classic bid-rent model. This short history sets the stage for exploring some of the ways in which communities beyond central cities are experiencing the same (or similar) conditions that have traditionally been attributed to "shrinking" cities.

A brief history of the American suburb

Suburbanization can be conceptualized as the outward (horizontal) expansion of built-up land relative to a central city. Consistent with the bid-rent model, suburbanization therefore involves increasing the intensity of land use at or beyond the urban fringe, specifically by developing previously open space. As the preceding section suggested, this conversion process predominantly results in the production of new *residential* spaces, at least during the early phases of suburbanization. In the United States, early suburban residential developments took on forms that reflected the pursuit of the **American Dream**: they tended to be communities of owner occupants living in detached, single-family homes on comparatively large lots, located in what were claimed to be quiet, safe neighborhoods separated from commercial and industrial land uses, and insulated from other urban push factors such as crime, substandard property conditions, and concentrated poverty.

Over time, however, suburbs began to evolve new forms, with some becoming cities unto themselves. As a consequence, suburbs are diverse, and this diversity of suburban forms and development patterns have made many of these communities vulnerable to the same changes that are observable in several central cities. The remainder of this section traces some of these suburban development patterns through time. In doing so, we adopt the following working definition: a **suburb** is a developed area outside a central city, either incorporated or unincorporated (see Ch. 8), that can feature a mix of land uses but is predominantly residential.

The walking city, pre-1815

Prior to the advent of the automobile and public transit, people generally walked to work. The **walking city** was the norm prior to 1815, and it had five distinctive characteristics. First, it was dense and congested. Densities soared as high as 75,000 people per square mile as people clamored to be as close to the CBD labor market as possible. Second, it was easily distinguished from the surrounding countryside. Often, there was a visual boundary such as the end of the pavement, or in the case of many European cities, a physical boundary such as a wall. Third, the walking city was a mixed-use conurbation that contained the totality of daily needs. Government offices and courts were interspersed with residential properties, commercial stores, and industrial facilities, with little delineation between land uses. Fourth, there was little to no inter-municipal commuting, as residents lived in close proximity to where they worked. Lastly, the most sought after and valuable land was in or near the CBD (refer to the preceding discussion of Alonso's bid-rent model). Before new transit technologies made it possible for residents to take on longer-distance commutes, even (and especially) the wealthiest households lived in the center city.

The early commuting city, 1815–1890

Beginning around 1815, U.S. metropolitan areas saw their first significant turn toward suburbanization. From this point, successive waves of transportation improvements – beginning with steam ferries, the horse-drawn omnibus, commuter railroads, and cable cars – increased the rate and magnitude of suburbanization by providing more people with lower-cost options for commuting to work.

The steam ferry had an immediate impact on suburbanization. The first continually scheduled ferry service began in 1814 in New York City, connecting Brooklyn to Manhattan. The low cost of commuting into Manhattan from outside boroughs, along with new home construction in less congested neighborhoods, helped Brooklyn attract a generally socially homogenous and wealthy population relative to the diverse population of Manhattan. Ferry services across the East River grew rapidly, and the number of daily crossings climbed into the thousands by the 1850s. Brooklyn's population (as King's County) grew from nearly 6,000 in 1800 to 138,000 by 1850, and reached more than a million by 1900. The steam ferry had similar effects on other U.S. cities, such as Boston in the East, Cincinnati in the Midwest, and Oakland in the West.

The steam engine's role in the industrialization of U.S. cities was connected directly to transportation, namely railroads and street cars, which were supporting increased and more dispersed suburbanization by the mid-1800s. Although some cities had local commuter train services as early as the 1830s, regional railroads did not provide access to areas farther from the city until the latter half of the nineteenth century. For instance, the commuter railroad was a dominant factor in the development of suburbs such as Llewellyn Park outside New York City. Construction for Llewellyn Park began in 1853 on 400 acres west of Manhattan. The original parcel was located within a mile of a railroad station that offered a thirteen-mile railroad commute to Manhattan. Llewellyn Park was developed as the romantic visualization of the country manor that ameliorated the trappings of city living with one- to ten-acre lots, more open space, and increased privacy. It also was one of the first planned communities in the United States, pioneering many of the characteristics evident in newer suburbs, including curvilinear streets, a break from the efficient grid pattern, and restrictions for residential uses only. The commuter railroad was dominant in the development of Chicago's suburbs as well. The city was a hub of economic activity connected to radiating spokes of railroad lines, along which suburbs sprouted up in the mid- to late-1800s, including Evanston, Aurora, Hyde Park, and Lake Forest, an exclusive suburban enclave of well-planned streets and finely designed houses. By the end of the 1880s, Chicago's suburban commuting population was more than 300,000.

Electric streetcars and the early automobile, 1890 to 1945

Up until the late 1880s, suburbanization was driven by newer transportation technologies, each successively opening up new areas outside the city for development – primarily for the wealthiest residents of America's cities. However, the development of the electric streetcar changed the urban/suburban landscape and began to make the suburbs accessible to the middle class.

The first reliable and successful electric streetcar system was designed and constructed in Richmond, Virginia, in the late 1880s. As it proved efficient and effective in moving commuters to and from the city, it was quickly replicated in other municipalities. By 1890, the streetcar had all but replaced the horse-drawn and steam-powered cable street car. With its larger passenger capacity and increased speeds, the electric street car could carry more people to farther areas more rapidly than steam engines. A direct consequence of this new, faster, and higher-capacity mode of transport was therefore more development farther from the central city (as predicted by the bid-rent model).

At roughly the same time as the electric streetcar was revolutionizing the urban commute, the next great turning point in suburban form was taking shape. Namely, the invention of the automobile near the end of the nineteenth century would ultimately take the magnitude, location, and form of suburbanization into previously unknown territory. Although still relatively unattainable for most Americans, by the early 1900s the automobile was already starting its meteoric rise to becoming American's preferred mode of transportation. From 1915 to 1925, the increasing affordability of the automobile pushed the number of registered cars in the United States from 2.3 million to 17.5 million. The relatively simple surfaces needed for cars (i.e., paved roads) compared to the infrastructure needed for rail travel meant that suburban development no longer needed to follow the linear patterns of rail lines and street car systems. Instead, the automobile offered a gateway to undeveloped land at the farthest reaches of urban commuting zones.

Despite the brave new suburban and rural worlds that were made accessible by the automobile, suburban development did slow during the 1930s as the United States experienced the Great Depression. During this time, unemployment, which affected nearly one out of every four Americans in 1932–1933, decreased the demand for new suburban homes. In addition, the rate of foreclosure on existing home mortgages was growing, resulting in, among other things, banks being less likely to initiate new mortgages. Automobile ownership also slowed during this period, but it did not cease altogether. That would change rapidly in the ensuing decades, when mass automobility became the norm.

Mass automobility part 1, 1945 to 1970

Robert Beauregard has referred to the period from 1945 to 1970 as "the Short American Century" – a time when the United States achieved global economic and political prominence, and when the population enjoyed plentiful job opportunities, rising wages, and general prosperity (Beauregard 2006).

Limited construction of new homes during the Great Depression left the United States with a temporary shortage of available housing at the end of World War II. With servicemen returning from the war and the economy ramping up, the confluence of plentiful jobs, rising wages, and low housing supplies was met by new mortgage policies from the U.S. federal government that were designed to insure mortgages and increase housing stocks. The Federal Housing Administration (FHA) and Veterans' Affairs (VA) mortgage programs supported a massive building boom that saw the number of new home starts increase from 114,000 in 1944 to 1.7 million by 1950. Homeownership rates, which were generally stable from

1900 to 1940, spiked from 43.6 percent in 1940 to 61.9 percent in 1960. It was at this point that the percentage of people living in cities began to decline. In fact, during the same time interval (from 1940 to 1960), the percentage of Americans living in suburbs doubled from 15 to 30 percent (Leyden and Goldberg 2015). These patterns of migration were largely enabled by massive federal government investments in the 1950s and 1960s toward establishing an autocentric transportation network consisting of state and interstate highways.

The result of these substantial housing policies and associated construction was a significant change in development patterns. The new development patterns exhibited five characteristics that were markedly different from earlier urban forms. First, new suburbs and subdivisions were geographically and economically disconnected from the central city. It was no longer necessary for residents and businesses to be close to rail or streetcar lines, as the freedom offered by the automobile – together with the new network of highways and roadways – opened up vast areas outside the city for people to enjoy suburban life. Second, density declined as individual lots became larger. Consider Brooklyn, where most brownstone rowhouses were 20 feet wide with lots only 100 feet deep – a typical lot size in the early twentieth century U.S. city. By comparison, new single-family homes in Levittown, NY, the self-anointed "First Suburb" of America, were being built on lots triple that size (60x100 feet).[2] Third, new building materials during this time period enabled homes to be mass produced. The result was large subdivisions of homes with little architectural variation. Fourth, homeownership became an affordable option for a much larger percentage of the population, due in part to rising wages, federally backed mortgages, and economies of scale in mass home production. Lastly, suburbs tended to be racially and economically homogenous (Jackson 1985). Low-income and minority populations – in many cases one and the same – were largely excluded from post-World War II suburbanization. This exclusion occurred through then-legal discriminatory lending practices, or, in the case of the builder of Levittown, NY, the outright refusal to sell to anyone who was not white. In fact, the deed signed by purchasers explicitly stated that the owners were not permitted to allow the property to be used or occupied by anyone who was not white.

The driving force behind this post-War suburbanization that disconnected the suburbs from the city was the automobile. The auto gave homeowners a vehicle from which to access almost any location in a given region, and homebuilders responded with new subdivisions on the urban fringe.

Mass automobility part 2, 1970 to present

As people relocated farther and farther from the city, commerce and industry responded similarly. Where CBDs of older industrial cities lost (generally manufacturing) jobs at alarming rates, the emergence of large retail malls and business parks beyond city limits gave way to suburban **edge cities**: business districts or suburban cores with considerable office and retail space, and, by extension, ample white-collar and service-based employment opportunities (Garreau 1991). As shopping centers and spaces of consumption, edge cities eventually became attractive to relatively low-skilled (and lower-income) workers. Namely, suburban employment centers offered a host of retail and hourly wage jobs. With low-skill employment opportunities drying up in center cities, many lower-income households – consistent with the logic of the *bid-rent model* – sought out jobs and housing in edge cities. Workers who found low-wage employment in edge cities could minimize their transportation costs by migrating to those edge cities, rather than facing the high commuting costs associated with remaining in center cities. At the same time, relatively wealthy households began to build newer housing on undeveloped land beyond the built-up areas of edge cities, thereby setting off further rounds of suburbanization.

With the formation of edge cities (i.e., suburban cores) and the corresponding emergence of new suburbs beyond the boundaries of edge cities, commuting patterns began shifting from suburb-to-city to suburb-to-suburb during the second phase of mass automobility. In 2000, for example, the highest percentage of commutes to work, about 1 in 3, were suburb to suburb (American Association of State Highway and Transportation Officials 2015). As suggested in the previous paragraph, these geographic changes to the metropolitan landscape coincided with major changes to urban economies. In the early 1970s, which marked the end of Beauregard's Short American Century, the United States was experiencing a deep economic recession. New waves of deindustrialization saw manufacturing jobs vanish not only from central cities but from the United States as a whole at massive scales. This sort of **economic restructuring** in America, which is tied to globalization of the world economy, has not played out evenly in all metropolitan regions. Instead, just as some (parts of) cities have shrunk and declined while others have not, many suburban spaces have experienced sharp shrinkage and decline – thereby going against popular perceptions of the wealthy and growing suburbs.

The next section unpacks some of the details of how *economic restructuring* in the United States has contributed to the emergence of shrinkage and decline beyond central cities. Prior to beginning that discussion, however, Figure 5.2 illustrates general (approximate) patterns of urban and suburban growth in the United States since 1850. The data come from the U.S. Census Public Use Microdata Sample (PUMS), in which survey respondents are categorized as living (1) within a metropolitan area or (2) not in a metropolitan area. Residents who live in metropolitan areas are further classified as living (a) within the principal city of a metropolitan area or (b) not within the principal city of the metropolitan area. Although metropolitan area (and city) boundaries may change from decade to decade, this generic classification scheme is consistently available in the PUMS data dating back to 1850.[3] Figure 5.2 therefore provides a useful (though approximate[4]) summary of the *trends* that have been articulated in this section.

Initially, during the *early commuting* phase of urbanization (1850–1890), principal city populations grew steadily, while "outside principal city" metropolitan populations – denoted in the graph as "suburbs" – remained essentially static and insubstantial. Beginning in 1890, as principal cities attracted large numbers of rural residents through the promise of CBD employment opportunities, some urban households took advantage of new transit modes (e.g., the electric streetcar)

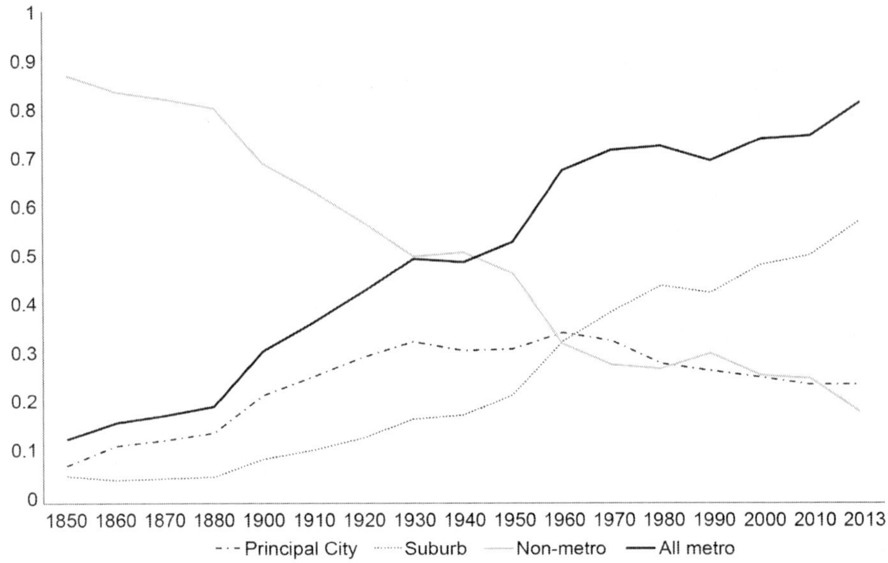

Figure 5.2 Estimated fraction of the U.S. population living in principal cities, suburbs, and non-metropolitan areas (Data source: http://ipums.org). All metro is combined principal city and suburb populations.

and relocated to the suburbs. Principal city and suburban growth then continued unabatedly through roughly the 1930s, when the economic hardships of the Great Depression placed new constraints on most households' geographic mobility. As World War II came to an end, principal city growth mostly stagnated while new federal housing policies and an economic upturn motivated residents to relocate to suburban communities at a rapid pace. Since the 1950s and 1960s, principal city growth has remained mostly flat or even ticked downward (especially in shrinking cities), while suburbanization has risen steadily. Overall, the population living in metropolitan areas (predominantly cities and suburbs) has far exceeded the population living in non-metropolitan (rural) areas (refer to the solid lines in Figure 5.2).

Local consequences of national economic restructuring and deindustrialization

The overall trends pictured in Figure 5.2 reflect three key shifts in U.S. economic activity over time. First, the overwhelming rural character of the American population through the mid- to late-1800s – i.e., through the periods of the *walking city* and the *early commuting city* – describe a national economy dependent on **primary sector industries**. Primary sector industries are those that rely on natural resource extraction and include activities such as agriculture and mining. Since the earliest days of European exploration, the United States has been viewed as a land of abundant natural resources. Plentiful supplies of cultivatable land and raw materials fueled much of the country's early economic development. Consequently, the nation's peoples tended to locate in close proximity to these predominantly rural resources. Second, from the early 1900s through World War II – the *electric streetcars and early automobile* period described above – America excelled at making "things". Technological and procedural innovations paved the way for mass-produced goods that could fill rising consumer demands quickly and relatively inexpensively. Although technology did allow for more automation in manufacturing processes during this time period, assembly lines and task specialization still required ample manpower. The promise of new, higher-paying jobs in these **secondary sector industries** (manufacturing and construction) therefore pulled residents from lower-paying jobs in rural communities into central cities. Urbanization occurred at a remarkable pace and scale, to the point where the average U.S. resident lived in an "urban" metropolitan area by 1920.

As cities became more congested, and the end of World War II brought about lower aggregate demand for manufactured products being made in the United States, another round of *economic restructuring* began. In this third shift, secondary sector jobs were relocated, first to more profitable parts of the United States (e.g., the South and West where many jobs were not unionized), and then to other nations where wages and production costs could be lowered even further in pursuit of higher profits. This *movement of capital assets across national borders* is a manifestation of *globalization*. In a very simple and broad sense, **globalization** refers to the increased integration of economic activity across national political borders, which results in increased international economic interdependence (Bluestone, Huff Stevenson, and Williams 2008). Globalization has taken place at lightning speeds since the end of the Short American Century, as advances in transportation and communication technologies drastically reduced the costs of moving and exchanging human, physical, intellectual, and financial capital assets between locations. Global businesses are now able to offshore their production processes to (often developing) countries with favorable regulations or low-cost labor supplies, all while managing logistics from headquarters in the United States and investing profits in Europe or the Caribbean. According to Friedman (2005), the world is flat(tening) – capital and cultural resources traverse space seemingly effortlessly in the era of globalization.

In the context of a globalized economy, the third shift in U.S. economic activity referenced above can be explained using the concept of *comparative advantage*. Economists use the term **comparative advantage** to refer to situations in which one entity (e.g., a country or region) can produce a particular good or service at a lower *relative cost* than a second entity. Relative cost in this sense is the quantity or value of goods or services forgone when a given labor supply produces a given type of commodity with a fixed set of inputs (also called *opportunity cost*). For example, if a given labor supply is capable of producing two different goods – say, beer and textbooks – then the relative cost of producing textbooks is the amount of beer that could have been produced by the labor supply using the same set of inputs. Assume that with the same set of inputs, region A can produce one textbook or three barrels of beer, and region B can produce two textbooks or four barrels of beer. In this case, region B enjoys an **absolute advantage** in producing both goods – it can produce more textbooks *and* more beer than region A using the same set of inputs.

Observe, however, that the *relative cost* of producing beer is higher for region B than for region A. Every barrel of beer produced by B costs B ½ of a textbook, while each barrel produced by A costs A only ⅓ of a textbook. In these terms, region A has a **comparative advantage** in beer production. Comparative advantage theory therefore recommends that region A should *specialize* in (i.e., use its inputs exclusively for) beer production, while region B should *specialize* in textbook production. The two regions can then trade with one another, so that each is able to obtain more of both goods than what would have possible under a regime in which both regions attempted to produce both goods "in house".

On that backdrop, it is often acknowledged that developed countries such as the United States have a comparative advantage in knowledge- and service-based (**tertiary**) industries. At the same time, developing countries tend to have a comparative advantage in *primary* and *secondary* industries. Differences in labor regulations, labor force education, cost of living, quality of life, and local purchasing power, among other things, generally allow firms to engage in primary and secondary economic activities at much lower costs in developing, as opposed to developed, nations. For these and other reasons, globalization has brought about new geographies of labor specialization. Labor intensive (especially secondary industry/manufacturing) jobs and production processes have and continue to be exported from the United States and other developed nations to developing countries; which has paved the way for developed nations to specialize, increasingly, in tertiary sector industries and establish knowledge- and service-based economies (Dicken 2007).

For the United States, this restructuring of the global economy has meant massive *deindustrialization*. Figure 5.3 depicts average annual data from the U.S.

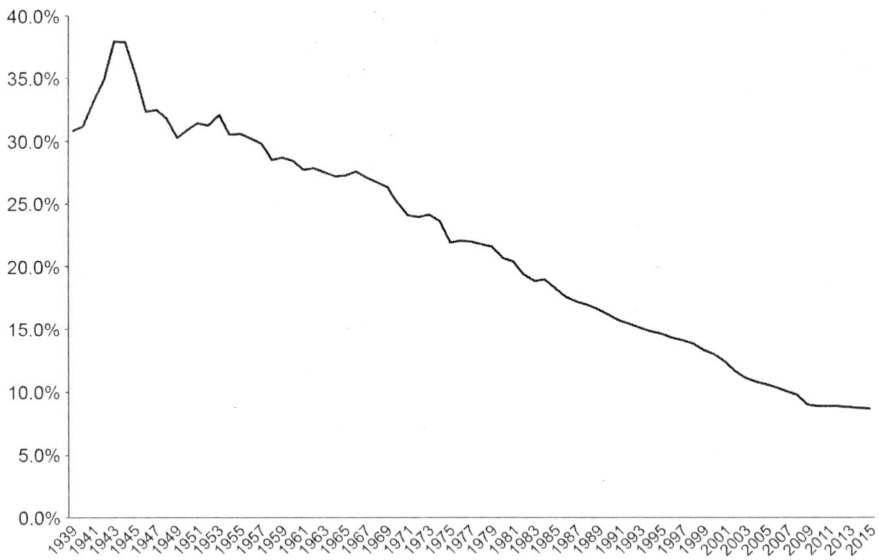

Figure 5.3 Manufacturing employees in the United States, by year, as a percentage of all non-farm workers (Data source: http://data.bls.gov/timeseries/CES3000000001)

Bureau of Labor Statistics (BLS) that quantifies the declining share of manufacturing jobs in the U.S. labor market since 1939. Specifically, the figure graphs the average annual number of manufacturing employees divided by the annual average number of total (non-farm) employees. The beginning of World War II in 1939 placed robust new demands on the U.S. economy to manufacture goods, and the percentage of non-farm production workers in the economy accordingly peaked during the war years at about 38 percent. With few exceptions, however, this value has been steadily declining since the end of the war in 1945. According to the BLS data, manufacturing employees currently account for less than 9 percent of all non-farm workers in America.

Regional filtering, diversifying suburbs

Deindustrialization substantially altered the U.S. economic and employment landscapes. Figure 5.4 combines the PUMS metropolitan population data from Figure 5.2 with the decadal BLS manufacturing data from Figure 5.3 to show how manufacturing job loss and suburban population growth appear to be move in opposite directions. Insofar as manufacturing operations historically concentrated in urban centers, the disappearance of these firms motivated workers to search for new employment opportunities elsewhere. This process led job seekers (and employers) to the growing retail and service sectors that were locating in suburban communities in general, and edge cities in particular. In addition to new

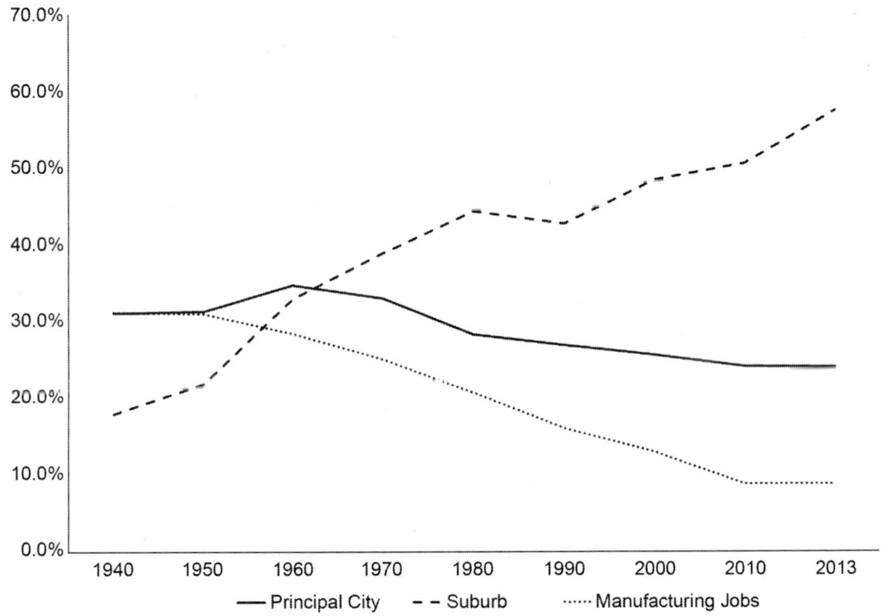

Figure 5.4 Manufacturing job loss compared to changes in the suburban and principal city share of U.S. population (Data sources: refer to Figs. 5.2–5.3)

jobs, housing opportunities also became increasingly accessible in the suburbs as deindustrialization progressed.

As discussed above, post-World War II federal policies (1) made homeownership more affordable for more people and (2) created vast automobile-oriented infrastructure networks to support new development and facilitate commuting beyond the urban fringe. In the shadows of both these federal policies and the macro-economic restructuring toward the tertiary sector, many urban dwellers were pulled to existing suburbs. As older suburbs grew in size and population, new suburban communities were established even farther from city centers, and the built landscape continued its horizontal/outward expansion.

This process resembled the *filtering* model of *intra-urban* change introduced in Chapter 2 and discussed in greater detail in Chapter 4. In this case, though, *intra-regional* filtering involved depreciation of homes and congestion in the older suburbs and construction of newer (typically larger) homes in newer, more distant suburbs. As relatively wealthy households left the older, so-called "first ring" suburbs for newer developments, their former (now depreciated) homes became affordable to relatively lower-income households. The result was a "downward succession" or "income filtering" process by which the physical and social fabrics of suburban communities underwent qualitative change – much like city neighborhoods in the urban filtering model (Ch. 4) – from the inside out.

While the idea that metropolitan regions experience succession and change from the inside out is vastly oversimplified – as were the *concentric ring* or *sector* models to which intra-city housing filtering was originally thought to apply (see Ch. 4) – it is a useful abstraction that speaks to the fact that "the suburbs" are not a homogeneous class of human settlements. Indeed, many suburban communities are now experiencing the same downward quantitative and qualitative adjustments that have traditionally been attributed to cities. Along these lines, notions that suburban communities have uniformly prospered while central cities have shrunk and declined are often unrealistic. Recent scholarship has challenged these notions through various attempts to highlight the diversity of suburban communities and articulate how (or if) those communities experience decline. For now, we turn to the longitudinal census tract-level dataset featured in Chapters 2–4 to empirically investigate the extent to which *actually existing* shrinkage and decline exist in non-central-city communities.

Patterns of non-central-city shrinkage and decline in the United States, 1970–2010

Chapters 2 and 3 introduced consistent and replicable approaches for identifying shrinkage and decline using empirical data from the Brown University LTDB (Logan, Xu, and Stults 2014). For *shrinkage*, we adopted the *threshold* proposed by Schilling and Logan (2008), by defining shrinking tracts to be those that lost 25 percent or more of their total population over a four-decade time horizon (from 1970–2010). From there, we computed the annual average [exponential] rate of population loss needed to achieve this forty-year threshold. This *threshold rate* can be used to identify tracts that are currently on pace to lose 25 percent or more of their total population in the next ten, twenty, or thirty years. Hence, based on observable population change data from the U.S. decennial census, one can determine which tracts have been shrinking – i.e., losing population at a rate that is greater than or equal to the threshold rate – since, for instance, the year 2000. Such tracts can be said to have endured relatively recent population *shrinkage*.

Along those same lines, to operationalize *decline* we developed a variable called the *G-index of concentrated disadvantage*. The G-index is a proxy measure for the degree of *concentrated disadvantage* (CD) in a geographic unit at a static point in time (Sec. 3.3). Because CD is an indicator of *distress*, comparing change in the G-index over time allows researchers to identify places that became *more distressed* – or, in other words, *declined* – between two given time periods (Fig. 3.1). In the analysis of decline in Chapter 3, we leveraged the G-index to identify all tracts that declined (i.e., experienced atypical increases in CD) during the same forty-year interval, from 1970 to 2010, for which we previously analyzed population shrinkage. Note, though, that because the G-index is derived from time-varying census data, we are also able to pinpoint tracts that have had more recent experiences with decline. In particular, with the Brown University LTDB data we can compute each tract's G-index in, say, 2000, and identify all tracts that have experienced atypical increases in CD since that time. Such tracts can be claimed to have endured relatively recent *decline*.

With these points in mind, it is possible to compare the distributions of census tracts that have undergone long-term (1970–2010) shrinkage, decline, or coupled shrinkage and decline, to the distributions of tracts that have experienced these phenomena more recently (2000–2010). Crucially, the Brown University LTDB (Logan, Xu, and Stults 2014) includes a dichotomous variable that indicates whether a given census tract is part of the core city in its metropolitan region. Using this variable, we can uncover the rough extent to which long-term patterns of shrinkage and decline, and/or more recent patterns of shrinkage and decline, do or do not concentrate in central cities.

It is important to once again note that the set of census tracts for which all relevant data are available for the full temporal extent of our analysis (1970–2010) does not cover the entire conterminous United States. Recall that the United States was not fully "tracted" in 1970, meaning that the whole country had not yet been divided into census tracts by the start of our study period (see Ch. 2). The parts of the country that were tracted, and which accordingly contain all relevant data, are primarily located within metropolitan regions (see Fig. 2.1). Thus, the results derived herein apply most readily to metropolitan areas, and additional work is needed to evaluate the extent to which our findings are applicable in non-metropolitan parts of the country.

In this subsection we address two primary questions:

1. Do central cities contain a disproportionate share of census tracts that shrank, declined, and/or both shrank and declined over the past forty years?
2. Do the distributions of shrinkage and decline between central cities and non-central cities appear to be shifting? More specifically: (a) are the census

tracts that experienced long-term shrinkage and/or decline (from 1970–2010) different from the census tracts that experienced more recent shrinkage and/or decline (from 2000–2010)? And, if so, (b) are these phenomena more or less concentrated within central cities?

The top portions of Tables 5.1 through 5.3 address question #1 by, respectively, showing how the tracts that experienced (i) population shrinkage (Fig. 2.2), (ii) decline (Fig. 3.2), and (iii) both shrinkage and decline (Fig. 4.1), from 1970–2010, are distributed between central cities and non-central cities. The contingency tables reveal that all three of these phenomena are disproportionately and significantly concentrated within core cities. Thus, shrinkage and decline have the appearance of being "city" or "urban" issues (though the fractions of shrinking, declining, and shrinking-declining tracts found in non-central cities are all significantly greater than zero).

Nevertheless, our efforts to answer question #2 suggest that these "urban" issues might be shifting away from central cities. The lower portions of Tables 5.1

Table 5.1 Distribution of tracts that shrank between 1970 and 2010 and tracts that shrank between 2000 and 2010, by their location

	In Central City	*Not In Central City*
Shrank (1970–2010)	5,136	2,250
Did Not Shrink (1970–2010)	19,158	25,835
» *Central cities account for 46.4% of all tracts and 69.5% of shrinking tracts*		
Shrank (2000–2010)	5,903	3,024
Did Not Shrink (2000–2010)	18,391	25,061
» *Central cities account for 46.4% of all tracts and 66.1% of shrinking tracts*		

Notes: Chi-squared tests on both contingency tables are significant with $p << 0.001$. Shrinkage is defined as an annual exponential rate of population loss of –0.719 percent or greater (see Chapter 2).

Table 5.2 Distribution of tracts that declined between 1970 and 2010 and tracts that declined between 2000 and 2010, by their location

	In Central City	*Not In Central City*
Declined (1970–2010)	5,653	2,480
Did Not Decline (1970–2010)	18,337	25,314
» *Central cities account for 46.3% of all tracts and 69.5% of declining tracts*		
Declined (2000–2010)	3,380	2,601
Did Not Decline (2000–2010)	20,624	25,293
» *Central cities account for 46.3% of all tracts and 56.5% of shrinking tracts*		

Notes: Chi-squared tests on both contingency tables are significant with $p << 0.001$. Decline is defined as a G-index increase of one or more standard deviations above the mean G-index change for a given time interval. The mean change in the G-index from 1970–2010 was 8.9 points with a standard deviation of 9.4. The mean change in the G-index from 2000–2010 was 1.6 points with a standard deviation of 5.4. Observations with missing data are excluded from the analysis.

Table 5.3 Distribution of tracts that shrank and declined between 1970 and 2010 and tracts that shrank and declined between 2000 and 2010, by their location

	In Central City	*Not In Central City*
Shrank and Declined (1970–2010)	1,547	302
All Other Tracts (1970–2010)	22,443	27,492
» *Central cities account for 46.3% of all tracts and 83.7% of declining tracts*		
Shrank and Declined (2000–2010)	1,003	390
All Other Tracts (2000–2010)	23,001	27,504
» *Central cities account for 46.3% of all tracts and 72.0% of shrinking tracts*		

Notes: Chi-squared tests on both contingency tables are significant with $p << 0.001$. See notes on Tables 5.1 and 5.2 for definitions of *shrinkage* and *decline*. Observations with missing data are excluded from the analysis.

through 5.3 show how tracts that have been shrinking, declining, and both shrinking and declining since 2000 are distributed between central cities and non-central-city locations. These tracts were identified using the approaches described in Section 5.6. The contingency tables reveal that, while more recent manifestations of all three phenomena are still significantly and disproportionately concentrated in central cities, the magnitude is decreasing in all three cases for central cities. In other words, (i) 66.1 percent of tracts that have been shrinking since 2000 are within central cities compared to 69.5 percent of tracts that shrank between 1970 and 2010; (ii) 56.5 percent of tracts that declined between 2000 and 2010 are within central cities compared to 69.5 percent of tracts that declined from 1970 to 2010; and (iii) 72.0 percent of tracts that shrank and declined between 2000 and 2010 are within central cities compared to 83.7 percent of tracts that shrank and declined from 1970 to 2010.

Tables 5.1 through 5.3 suggest not only that shrinkage and decline are more than "urban" issues but also that patterns of these phenomena might be de-concentrating. More precisely, shrinkage and decline are seemingly becoming more prevalent in places beyond core/central cities.

To test whether this implied tendency is statistically significant, Tables 5.4 through 5.6 present the results of two-sample **McNemar tests** (Adedokun and Burgess 2011). McNemar tests can detect significance in patterns of census tracts whose statuses as shrinking, declining, both, or neither, "change" over time. For example, it is possible for a tract that experienced long-term shrinkage (from 1970–2010) to have had a relatively stable population from 2000–2010. Thus, while the tract might be classified as *shrinking* based on forty-year observations from 1970–2010; if we are interested in *recent* events, then we would probably classify the same tract as *not shrinking* based on ten-year observations from 2000–2010. McNemar tests make these sorts of comparisons over an entire dataset using contingency tables that compare tracts' classifications for one time period (rows) to their corresponding classifications at another time period (columns). Two-sample McNemar tests go a step further to compare how inter-temporal changes observed

Table 5.4 Two-sample McNemar test for differences in changing patterns of shrinkage between tracts inside and outside central cities

	In Central City (% of column total)	*Not In Central City (% of column total)*
Tract Tended Away from Shrinkage	2,208(42.6)	1,314(38.6)
Tract Tended Toward Shrinkage	**2,975(57.4)**	**2,088(61.4)**
All Changing Tracts	5,183	3,402

Notes: $\chi^2[1] = 13.3; p < 0.001$. **Bold text** indicates the stronger direction of change in each tract type.

Table 5.5 Two-sample McNemar test for differences in changing patterns of decline between tracts inside and outside central cities

	In Central City (% of column total)	*Not In Central City (% of column total)*
Tract Tended Away from Decline	**3,444(74.7)**	1,406(48.0)
Tract Tended Toward Decline	1,166(25.3)	**1,523(52.0)**
All Changing Tracts	4,610	2,929

Notes: $\chi^2[1] = 555.5; p << 0.001$. **Bold text** indicates the stronger direction of change in each tract type.

Table 5.6 Two-sample McNemar test for differences in changing patterns of coupled shrinkage and decline between tracts inside and outside central cities

	In Central City (% of column total)	*Not In Central City (% of column total)*
Tract Tended Away from Shrinkage and Decline	**1,054(67.4)**	207(41.2)
Tract Tended Toward Shrinkage and Decline	510(32.6)	**297(58.8)**
All Changing Tracts	1,564	502

Notes: $\chi^2[1] = 108.2; p << 0.001$. **Bold text** indicates the stronger direction of change in each tract type.

in one sample (here, central cities) differ from the changes observed in another sample (non-central cities).

A more detailed description of the nature of the two-sample McNemar test along with a step-by-step illustration of how it was carried out is provided in Appendix B. For now, it is sufficient to say that the test operates by singling out all census tracts whose classifications as shrinking, declining, both, or neither "changed" for the two time periods analyzed above (i.e., the full period from 1970–2010 and the recent period from 2000–2010). For each phenomenon of interest (shrinkage, decline, and coupled shrinkage and decline), the test then breaks out the pool of all "status changing" tracts into four categories: (1) central city tracts that tended away from the phenomenon (e.g., tracts that endured population shrinkage from 1970–2010 but did not meet the threshold for shrinkage from 2000–2010); (2) central city tracts that tended toward the phenomenon; (3) non-central-city tracts that tended away from the phenomenon; and (4) non-central-city tracts that tended toward the phenomenon. Given the circumstantial evidence contained in Tables 5.1–5.3, a higher proportion of non-central-city tracts could tend *toward* shrinkage and decline relative to central city tracts.

Indeed, these expectations are supported by the three McNemar tests (Tables 5.4–5.6). Among tracts whose long-term (1970–2010) experiences with population shrinkage differed from their more recent (2000–2010) bouts with this phenomenon, the predominant tendency was to move toward shrinkage. Moreover, the tendency toward shrinkage was statistically significantly stronger in non-central-city locations compared to central cities (Table 5.4). These patterns of change are even more pronounced for cases of decline and coupled shrinkage-decline. Concerning the former, among central city tracts whose decline statuses "changed" for the two time periods under investigation, the overwhelming majority tended away from decline. The upshot is that conditions of central city decline might be stabilizing or improving in our sample of census tracts (refer to Appendix B). In contrast, the majority of status-changing tracts outside central cities tended toward decline – indicating that patterns of decline appear to be rapidly evolving as well (Table 5.5). As Table 5.6 shows, these same conclusions were found for patterns of coupled shrinkage and decline.

The findings from Tables 5.1–5.6 imply that shrinkage and decline should not be considered, and likely never were, exclusively urban issues. Both phenomena are increasingly affecting communities outside core cities. For instance, at the same time *decline* is starting to display inchoate signs of stabilizing or reversing in core cities, the phenomenon is on the rise in other parts of metropolitan regions. Because it is somewhat impractical and prohibitive to map these fine-grain pattern shifts across a national extent, the next section engages more directly with these emerging geographies of *suburban decline* through a short case study. The case study augments the results from above with a visual analysis that illustrates the *regional* – as opposed to the strictly urban – nature and character of shrinkage and decline.

Exploring suburban decline: St. Louis, Missouri

This section offers a glimpse into the changing patterns of *decline* in a single metropolitan region. The point of the exercise is twofold. First, it localizes the prior national-scale analyses to provide readers with a simple exploratory approach for monitoring decline within metropolitan areas. Second, using the Greater St. Louis region as a case, it supports our claims that decline is de-concentrating: where conditions of socioeconomic decline were once seen almost exclusively as urban issues, they are now spilling over into, or perhaps independently arising within, communities outside principal (in the case of St. Louis, shrinking) cities. In effect, then, the St. Louis case serves as an illustrative opportunity for us to visualize an example of the changing geographies of decline on a map.

St. Louis, Missouri, is a former transportation hub located on the Mississippi River in the Midwest United States, on the border of Missouri and Illinois, and across from the smaller city of East St. Louis, Illinois. Given its locational advantages with respect to water and rail transport routes, the city grew rapidly in the late nineteenth century, becoming one of the nation's premier manufacturing centers. The population of St. Louis exploded from 16,469 people in 1840, to 77,860 people in 1850, and 821,960 people by 1930. After a slight drop-off in population between 1930 and 1940, the city reached a peak of 856,796 people in 1950. Since that year, population has dropped continuously, to the point where the city reported only 319,294 residents in the 2010 U.S. decennial census. Meanwhile, the population of the Greater St. Louis metropolitan area has grown in all but one decade (1970–80) since 1930. During that time span, the number of people in Greater St. Louis more than doubled, from 1.14 million in 1930 to 2.79 million in 2010. The picture painted by these numbers seems to be one of a shrinking and declining central city surrounded by growing and thriving suburbs.

Nonetheless, pictures painted by aggregate data can often be misleading. To challenge the image of a suburban St. Louis insulated from the principal city's "urban problems", decline is evaluated here using the *G-index of concentrated disadvantage* (CD). Specifically, we extracted all census tracts from the Brown University LTDB – for which forty years' worth of data are available – that fall within the U.S. Census Bureau's current definition of the Greater St. Louis metropolitan statistical area (n = 571). Figure 5.5(a) maps the distribution of all tracts from this sample for which the G-index in 1970 was one or more standard deviations above the mean 1970 G-index for the Greater St. Louis region as a whole. With almost no exceptions, these "high CD" or *distressed* census tracts lie entirely within either the city of St. Louis (dark outline immediately to the west of the pictured Missouri-Illinois state boundary), or the city of East St. Louis (dark outline immediately to the east of the state boundary).

The implication is that, in 1970, high-concentrated disadvantage was principally an urban problem in the selected metropolitan region. By comparison, Figure 5.5(b) depicts all tracts for which the 2010 G-index was one or more standard deviations above the mean 2010 G-index in Greater St. Louis. Noticeably, high CD (distress) has penetrated the boundaries of St. Louis and East St. Louis, and is now a prominent feature in many adjacent ("first-ring") suburban communities.

The preceding visual analysis for Greater St. Louis demonstrates not only that suburban communities are subject to decline, as captured by the changing distribution of census tracts characterized by high-concentrated disadvantage (Fig. 5.5); it further suggests that the geographies of suburban decline might be at least partially consistent with expectations from the *filtering* model and related theories of

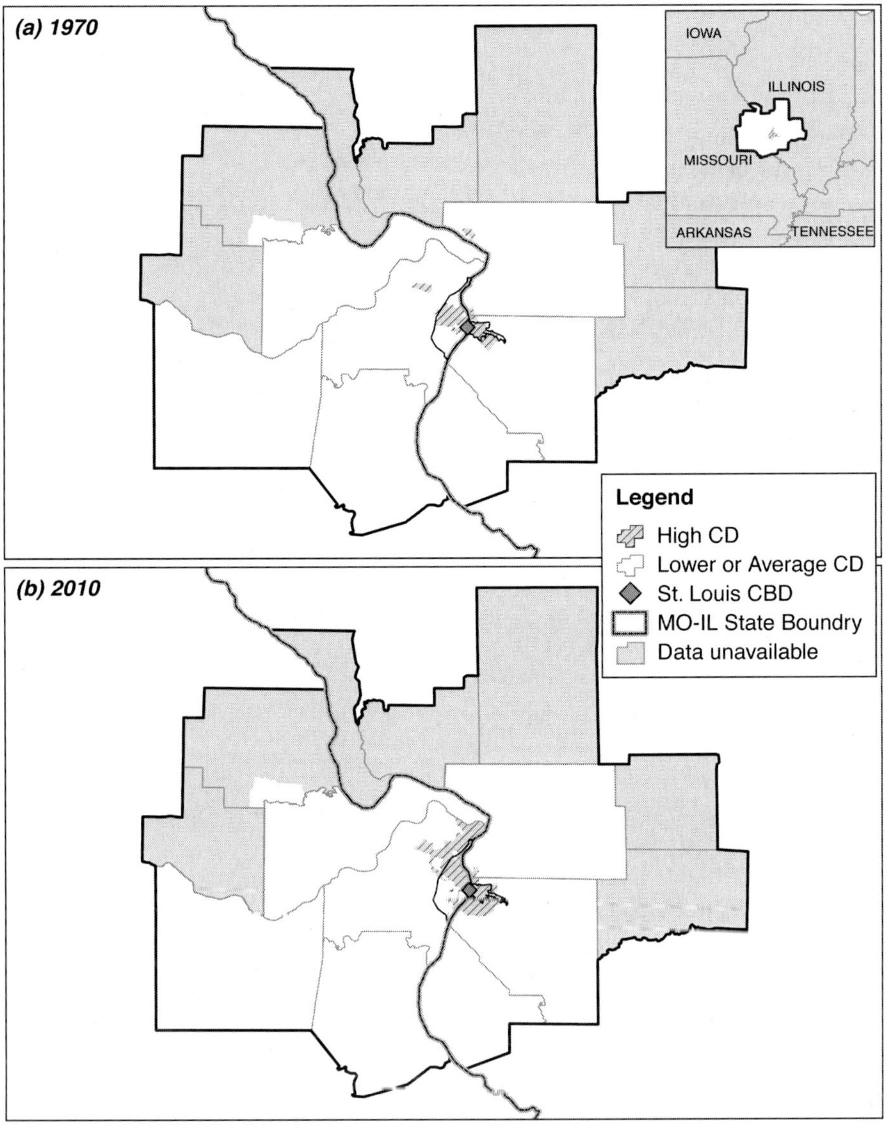

Figure 5.5 An exploratory analysis of regional decline in the St. Louis Metropolitan Statistical Area, 1970–2010 (Data source: Brown University LTDB)

urban decline (Ch. 4). Observe from Figure 5.5 that conditions of atypically high-concentrated disadvantage seemed to spread outward, or spillover, from the central cities to connected territories over time. If this increase in distress is interpreted as *decline* (refer to Fig. 3.1), then patterns of decline appear to be following the initial patterns of (sub)urban growth that were explicated in the context of the *bid-rent model* at the outset of this chapter.

Recent shrinkage, decline, and variation in social capital beyond central cities

Of course, just as suburbanization, associated patterns of filtering, and deindustrialization are not sufficient explanations for urban shrinkage and decline in general (Ch. 4), nor can these forces be fully responsible for producing patterns of *suburban* shrinkage and decline. Rather, we contend that the complex, context-dependent conditions that lead to shrinkage and decline within cities are also operating outside cities. The theoretical survey and analyses from Chapter 4 offered persuasive justification that *social capital* is at least one important condition. Here, we replicate the analyses of social capital indicators that were presented in Table 4.3, for *non-central-city tracts that have had recent experiences with shrinkage or decline.* In other words, the universe of analysis in this section is the 3,024 non-central-city census tracts identified in Table 5.1 that lost population between 2000 and 2010 at an annual rate that exceeded the *threshold* for shrinkage. Within this set, 2,870 tracts (95 percent) included data for all of the variables needed for the analysis (i.e., all variables used to construct the G-index to identify decline and all social capital indicators).

Selection of this set of census tracts is motivated by a desire to see if the patterns of differences that were observed in social capital variables in conjunction with *long-term urban shrinkage and decline* (Table 4.3) hold for situations of *contemporary suburban shrinkage and decline.* If similar patterns of differences in social capital can in fact be observed for these two seemingly different scenarios – i.e., long-term urban shrinkage/decline versus recent suburban shrinkage/decline – then it seems reasonable to make the relatively broad conclusion that social capital is very likely part of the conditions that associate with patterns of shrinkage and decline *in general.*

Table 5.7 presents the results from applying the Wilcoxon signed-rank tests from Table 4.3 to the set of census tracts described above. For more information on the choice of indicator variables, data, methods, and expected differences between the indicators, refer to Chapter 4. What matters most here is that similar patterns of results that were observed for the case of long-term urban shrinkage and decline in Table 4.3 are also observed in Table 5.7. Specifically, compared to their declining counterparts, non-central-city census tracts that (1) experienced recent population shrinkage but (2) did not endure significant qualitative decline during this period of population shrinkage seem to have:

- a relatively high number of civic organizations;
- a relatively high number of government offices;
- a greater share of long-term stakeholders;
- a greater predisposition to trust others in transactions characterized by asymmetric information;
- slightly lower income inequality; and
- greater internal community homogeneity.

Table 5.7 Wilcoxon signed-rank tests for equality of medians in independent samples

Variable	*Shrink Only*	*Shrink and Decline*	*Difference*	*Indicator of:*
Civic Organizations (per 1,000 persons per square mile)	3.5	2.4	1.1***	Bonding Connection
Government Offices (per 1,000 persons per square mile)	0.6	0.4	0.2***	Bridging Connection
Long-Term Owners (% of occupied units)	20.1	16.6	3.5***	Collective Norms
Used-Good Expenditures Per Capita (US$)	28.0	20.4	7.6***	Generalized Trust
Income Inequality Index	41.0	42.5	–1.5***	Social Cohesion (–)
Ethnic Diversity Index	37.9	50.5	–12.6***	Homogeneity (–)
n[a]	2,482	388		

[a] Excludes 154 cases with missing data; ***$p < 0.001$; universe: census tracts outside central cities that met the threshold for population shrinkage from 2000–2010 (n = 2,870 cases with complete data)

Concluding remarks

Understanding patterns of shrinkage and decline beyond central city borders is as complex as (if not more complex than) understanding how these processes play out inside individual cities (see Ch. 4). Consequently, the discussions and supporting illustrations and analyses presented in this chapter are far from an exhaustive account of suburban change. However, the synthesis of these materials suggests that shrinkage and decline are not localized phenomena that affect cities individually and only within their political borders. These problems are not "city problems" that can be ignored by other municipalities in a region. They are regional issues that likely require cooperative regional responses. Nevertheless, intra-regional cooperation – especially between suburban municipalities and core cities – is a difficult outcome to achieve in regions affected by shrinkage, decline, or both. On that note, the next four chapters focus on existing policies, new approaches, governance patterns, and a discussion of urban sustainability/resilience.

Notes

1 The academic literature on "suburban decline" has expanded rapidly in recent decades. Thus, the antiquated view of a shrinking-declining inner city surrounded by thriving suburbs is no longer a mainstay of planning scholarship (e.g., Short, Hanlon, and Vicino 2007). Such a view does, however, still underlie many popular perceptions of U.S. metropolitan regions (e.g., Favro 2010).

2 Interestingly, the original 6,000-square-foot lots in Levittown, NY pale in comparison to the average lot size of new single family homes currently sold in the US, which was 15,456 square feet in 2013.

3 NB: Figure 5.2 excludes the data categories "Not Identified" and "Unknown".
4 Note that the PUMS dataset does not include a "suburban" geographic category. As a consequence, the "living within a metropolitan area, outside the principal city" category described in this paragraph is used as a proxy for suburban population.

References

Adedokun, Omolola A., and Wilella D. Burgess. 2011. "Analysis of Paired Dichotomous Data: A Gentle Introduction to the McNemar Test in SPSS." *Journal of MultiDisciplinary Evaluation* 8 (17):125–131.

Alonso, William. 1964. *Location and Land Use: Toward a General Theory of Land Rent*. Cambridge, MA: Harvard University Press.

American Association of State Highway and Transportation Officials. 2015. *Commuting in America 2013: The National Report on Commuting Patterns and Trends*. Washington, DC: American Association of State Highway and Transportation Officials.

Beauregard, Robert A. 2006. *When America Became Suburban*. Minneapolis: University of Minnesota Press.

Bluestone, Barry, Mary Huff Stevenson, and Russell Williams. 2008. *The Urban Experience: Economics, Society, and Public Policy*. Oxford: Oxford University Press.

Dicken, Peter. 2007. *Global Shift: Mapping the Changing Contours of the World Economy*. 6th ed. New York: The Guilford Press.

Favro, Tony. 2010. "American Cities Seek to Discover Their Right Size." *City Mayors*. http://www.citymayors.com/development/us-rightsizing-cities.html

Friedman, Thomas L. 2005. *The World Is Flat: A Brief History of the Twenty-First Century*. New York, NY: Farrar, Straus & Giroux.

Garreau, Joel. 1991. *Edge Cities: Life on the New Frontier*. New York: Doubleday.

Großmann, Katrin, Marco Bontje, Annegret Haase, and Vlad Mykhnenko. 2013. "Shrinking Cities: Notes for the Further Research Agenda." *Cities* 35:221–225.

Jackson, Keith T. 1985. *Crabgrass Frontier: The Suburbanization of the United States*. New York: Oxford University Press.

Jessen, Johann. 2012. "Conceptualizing Shrinking Cities – A Challenge for Planning Theory." In *Parallel Patterns of Shrinking Cities and Urban Growth, Spatial Planning for Sustainable Development of City Regions and Rural Areas*, edited by Robin Ganser and Rocky Piro, 45–58. Burlington: Ashgate Publishing Ltd.

Leyden, Kevin M., and Abraham Goldberg. 2015. "The Built Environment of Communities and Social Capital." In *Social Capital at the Community Level: An Applied Interdisciplinary Perspective*, edited by John M. Halstead and Steven C. Deller, 31. New York: Routledge.

Logan, John R., Zengwang Xu, and Brian J. Stults. 2014. "Interpolating U.S. Decennial Census Tract Data from as Early as 1970 to 2010: A Longitudinal Tract Database." *The Professional Geographer* 66 (3):412–420.

Schilling, Joseph, and Jonathan Logan. 2008. "Greening the Rust Belt: A Green Infrastructure Model for Right Sizing America's Shrinking Cities." *Journal of the American Planning Association* 74 (4):451–466.

Short, John Rennie, Bernadette Hanlon, and Thomas J. Vicino. 2007. "The Decline of Inner Suburbs: The New Suburban Gothic in the United States." *Geography Compass* 1 (3):641–656.

6 Pro-growth urban policy

So far in this book we have examined various *patterns of* and *trends* in shrinkage and decline in the United States (Ch. 2–3), and we have discussed several factors that associate with these patterns through indicators and proxies of processes and conditions (Ch. 4–5). One broad takeaway from these efforts is that the fates of all places are different. Some places shrink while others grow. Some decline while others prosper. Still others experience reinforcing processes of coupled shrinkage *and* decline, often within "downward spirals" from which escape seems impossible (Emery and Flora 2006). Robert Beauregard has referred to these patterns of development, whereby some places thrive as others are left behind, as *parasitic urbanization* (Ch. 1). Whereas urbanization was once thought to bring growth to all cities in a relatively *distributive* manner, contemporary parasitic urbanization is associated with "winners" and "losers" in competitions for residents, economic investments, and other forms of capital (Beauregard 2006; Logan and Molotch 1987). As a starting point, it is useful to note that one belief still alive and well in many policy arenas is that parasitic urbanization might be a temporary departure from an equilibrium, or normal condition, of more distributive urban growth (see the discussion by Großman et al. 2013). For instance, consider that cities where population shrinkage has taken a firm hold are often those that experienced growth in the past, perhaps as a result of a manufacturing boom, which could not be sustained in an era of globalization. In such cases, returning to prior levels of population and economic activity is highly improbable (Leo and Anderson 2006). Nevertheless, urban planning and policy instruments in such cities remain overwhelmingly targeted toward these very scenarios. Put differently, there is a belief that shrinkage will eventually be followed by a return to growth (Hackworth 2014).

The next section reviews this *pro-growth* approach within the context of *neoliberal* ideology and *neoclassical economic theory*. Crucially, the worldview formed at the intersection of these paradigms predominates in, and simultaneously structures, contemporary political and policymaking arenas in Western cities (Peck and Tickell 2002). Engaging with this worldview therefore facilitates a discussion of the prevailing development logic in U.S. municipalities that Kantor (2010) calls the *American model of urban development*. Subsequent to the discussion of this American model, the chapter highlights selected policy instruments that are repeatedly deployed in shrinking cities to reverse their trajectories. This exercise is intended to illuminate the context of public policies. In doing so, we draw on numerous practical examples, as well as empirical data, from shrinking cities in the United States. Finally, the chapter concludes by summarizing recent calls for decline-oriented development logics to supplant growth-first mentalities in shrinking cities. This topic is continued in Chapter 7.

Neoliberalism and neoclassical economics

To better understand the thought process that underlies the development of many of the conventional policy tools used to combat shrinkage and decline in U.S. cities, it is necessary to grapple with both the values of, and theoretical propositions used by, proponents to rationalize those policies. That being said, a full treatment of these ideas requires its own volume. Here, we paint the requisite picture in relatively broad strokes.

Neoliberalism is construed as a doctrine (Glassman 2009) or an ideology (Brenner and Theodore 2002) that is predicated, minimally, on the following beliefs:

1 that open and competitive markets, which are characterized by capital mobility and well-defined property rights but are otherwise unregulated, are the ideal mechanisms for allocating resources in a society;
2 that such market-based resource allocation leads to optimal socioeconomic development outcomes, wherein aggregate gains from economic (growth) activities are maximized;
3 that individuals are both free, and have the individual responsibility, to make claims on these aggregate gains in their roles as entrepreneurs operating within the bounds and rules of the market; and
4 that the public sector's role in economic development should be limited to protecting property rights and supporting the creation and functioning of private markets.

With respect to the latter of these points, more than simply advocating for a limited role for government, many strict adherents to neoliberal ideology propound an all-out moratorium on collectivist policies (Peck and Tickell 2002: 381). In place of such [redistributive] actions, they suggest that, similar to their views on the roles of individuals, governments should behave as entrepreneurs that compete in market-like systems for residents and economic development projects (Peck, Theodore, and Brenner 2009). Skilled government-entrepreneurs will attract external economic investment, which, in turn, can draw in new residents (and wealth) and make all of a city's citizens better off in the process.

Competition is therefore a means for city-level economic accumulation. The more competitive a city, the larger will be its economy. In this way, as the term implies, the rewards from 'competition' (namely, economic development) are institutionalized metrics of *success*. It follows that to shrink is to *fail*. Hence, shrinking places, in attempts to improve their relative positions on this success-failure continuum, tend to put economic expansion at the forefront of their policy agendas (Leo and Anderson 2006). That is, because of the stigma associated with urban decline, and the attendant lack of established planning tools for managing decline (Dewar, Kelly, and Morrison 2012), even shrinking cities exhibit a *pro-growth* approach in the ways they plan for and govern urban change.

Together, the preceding ideas form much of the foundation of neoliberal ideology. This ideology gained significant momentum in the wake of economic and fiscal crises in the 1970s, as the crises were framed as failures of the "big government" spending programs from earlier decades. Subsequently, a *roll-back* version of neoliberalism swept through the United States (under President Reagan) and the United Kingdom (under Prime Minister Thatcher), where investments into social welfare programs were regularly substituted for market-based policies that sought to make individuals personally responsible for their well-being. This roll-back project would later be accompanied by a *roll out* version of neoliberalism that established new institutions for addressing social problems through market or market-like mechanisms.[1]

The steady rise in neoliberalism's popularity since the 1970s has meant that the foregoing beliefs now influence public policies at all levels of government in the United States, including the local. Keep in mind, though, that these are normative ideas. That is, they are beliefs – about optimal mechanisms of macro-allocation; about the impartiality and inclusiveness of markets; and so forth. Rarely do normative beliefs materialize from nothing. In the present case, the tenets of neoliberalism are inextricably linked to propositions from **neoclassical economic theory**. Whereas neoliberalism is understood as an ideology, and therefore contains statements about how the world *ought to* work (see the beliefs enumerated above), neoclassical economics is a school of thought about how the world *does* work. It strives to be a positive science, and, like all scientific theories, it makes several assumptions and abstractions. The following theoretical propositions from neoclassical economics capture fundamental simplifying assumptions and implications that feed into the neoliberal belief system.

- Economic agents (e.g., individuals, firms, and governments) are *rational*, *informed*, and *self-interested* decision-makers. They are aware of the costs and benefits of alternative choices, and they choose only those strategies that *maximize their individual well-being.*
- In the absence of high *transaction costs* and the presence of well-defined *property rights*, market exchange produces an *efficient* allocation of resources. More precisely, given a distribution of resources across a population of rational agents, competitive exchange in unregulated markets moves society (as if being led by an *invisible hand*) to an outcome in which no one agent can be made better off without reducing the welfare of another agent.
- This unregulated, "invisible hand" style of social organization works because relative prices convey information. Rational responses by economic agents to market signals (i.e., prices) are *equilibrating*. Price adjustments move

> markets to outcomes wherein the demand for and supply of goods and services are equalized. At such outcomes, no rational agent has an incentive to unilaterally deviate from his or her equilibrium strategy: all producers who wish to sell goods and services at a going market price are doing so; and likewise for all consumers who wish to purchase goods and services at that price.

The upshot is that the assumptions of individual self-interest and rational choice from neoclassical economics have several logical, theoretically grounded consequences that, if true, are quite appealing. Above all else, because rational actors only make mutually beneficial transactions with one another (hence their rationality), allowing those actors to have control over their own decisions in free market environments is the least costly means for ensuring that all welfare-enhancing transactions in a society are made. Put differently, whereas a central controller or government would need to acquire perfect information about the personal (private) preferences of all economic agents if it were to facilitate all "mutually beneficial" transactions in a society, markets perform this function seemingly automatically. Consequently, neoliberal thinkers, who tend to accept the individual rational choice model as a useful representation of reality, hold up the market as the ideal mechanism for engaging in socio-economic development. The influence of this above ideology is readily apparent in the prevailing American approach to urban development and the policies that flow from it.

The American model of urban development

While the following conditions are not specific to the United States, the American metropolitan landscape is recognized for its local, non-coordinated, and non-uniform land use controls, which are overlaid onto patchy, decentralized patterns of settlement (Dewar and Thomas 2012). Kantor (1995) observes that what does distinguish the United States from other developed nations, however, is that its decentralized settlement patterns are generally combined with overt resistance to higher levels of government. This resistance is arguably a manifestation of American neoliberal thinking. Specifically, from any given municipality's standpoint, horizontal cooperation with another municipality, under the (vertical) coordination of a regional institution, is individually costly and inhibits competition. As a result, regional solutions are likely to be perceived by neoliberal thinkers as "anti-market" (Hackworth 2014: 4). The particular development logic that is formed from these conditions and beliefs – the **American model of urban development** – is held up by three key pillars: devolution, inter-governmental competition, and public entrepreneurship.

Devolution

The notion that local government is better attuned to local needs than, and can simultaneously constrain the powers of, a central authority extends throughout

American history (Wood 1992). This dual confidence in the local and suspicion of the federal led to a preference for decentralist urban policy, which has framed America's response to post-industrial change (Kantor 2010: 5). In short, deindustrialization of America's urban manufacturing centers motivated many geographically mobile households to relocate from central cities to the urban fringe. Federal push factors incentivized out-migration, as national policies set up homeownership inducements and infrastructure (e.g., highways) that connected urban households with larger lots and newer housing opportunities farther from the city center (Ch. 5). What is more, state-level policy frameworks enabled fleeing households to incorporate into autonomous municipalities and set up their own local governments (Ch. 7). Through land use controls, these new communities were effectively able to insulate themselves from what they saw to be undesirable features of the declining urban landscape.

Over time, this process brought about a sort of suburban explosion, with numerous [at least initially] thriving suburban towns encircling a rapidly depopulating central city. The resultant fragmented political landscape offered geographically mobile households the opportunity to self-sort into communities according to the packages of local laws and public goods that they most preferred. Put another way, households could move to independent towns whose newly established land use controls and property tax systems made it possible to consume relatively more (and newer) housing services farther from the conditions of poverty and blight that were being left behind, and accumulating, in the city (Kantor 2010). This tendency to "vote with one's feet" is commonly known as the **Tiebout hypothesis**, named for the economist who popularized it in the 1950s (Tiebout 1956).

The fragmented migration patterns associated with the Tiebout hypothesis – and with suburbanization – were accompanied by neoliberal preferences for a "hands off" federal government approach to (sub)urban development. Whereas early and mid-twentieth century American federal policy played a significant role in shaping cities and funding urban renewal programs, expensive federal city-building programs became unattractive to taxpayers in the suburbs. Accordingly, independent jurisdictions pushed for greater autonomy in local growth (decline) management. Moreover, the documented failures of the nation's top-down urban renewal programs of the 1950s and 1960s buttressed neoliberal arguments that federal intervention into local affairs is harmful (Schilling and Mallach 2012). These sentiments led to a downward shift in governmental responsibilities, reflecting the political belief that individual municipalities benefit from permitting neoliberal forces, such as market-based competition, to guide changes in development (Kantor 2010: 6).

Inter-governmental competition

Together, rapid metropolitan fragmentation and devolution of growth (decline) management policy to the local level formed a sort of spatial marketplace. Like geographically mobile households in the Tiebout model, businesses (and private capital in general) are increasingly mobile in post-industrial cities. As former manufacturing cities transition from production to knowledge- and service-based economies, their corporate residents are no longer characterized by spatial fixity. Instead, corporate actors choose to locate in areas where land use controls, tax structures/incentives, and cultural and physical geographies make doing business most profitable for their companies and stakeholders. This reality can have the perverse effects of local environmental deregulation (Peck and Tickell 2002), public "giveaways" (i.e., large subsidies; see Weber 2002), and any number of other actions that effectively subordinate municipal authorities – and, by extension, their citizens – to the private sector (see Logan and Molotch 2007).

Nonetheless, viewed through the lens of neoclassical economics, this sort of "race to the bottom" is a rational reaction to existing circumstances. Devolved decision-making in a fragmented metropolis promotes inter-governmental competition. Under the given rules, cities that do not attract private development are branded as "losers", are perceived negatively from the outside, and develop an internal sense of powerlessness that feeds back to reinforce negative external perceptions (Leo and Anderson 2006). Consequently, even in shrinking cities that are unlikely to re-grow, public officials remain committed to pro-growth strategies, especially those aimed at attracting large (e.g., multinational) private businesses (Schatz 2012).

Public entrepreneurship

As intimated in the preceding section, local and state governments "routinely subsidize urban business investment with . . . tax, loan, regulatory, and other business incentive programs in an effort to bend the pressures of the marketplace to their favor" (Kantor 2010: 7). While these entrepreneurial programs take on many forms, most of which are heralded as *public-private partnerships*, they all point to the inseparability of the public and the private in the American model of urban development (MacLeod 2002: 256). Explicitly, the "role of public authorities has . . . changed from those who produce [metropolitan space] to those who promote and regulate its production" (Madanipour 2006: 181). That is, where public actors were once key place-*makers*, in the post-industrial, fragmented, pluralistic metropolis they have become place-*marketers*, whose focus has shifted from internal quality of life to external perceptions of businesses, tourists, and other holders of mobile capital (Blomley 2004).

The commitment to this entrepreneurial approach to urbanism has echoed throughout the halls of American government, as even the U.S. Supreme Court ruled that municipalities may seize property to sell to private developers for the purpose of local economic growth (*Kelo v. City of New London* 2005). In light of this ruling, arguments that a "growth-first" paradigm of urban development is normative in Western, particularly American, cities are all the more compelling (Peck and Tickell 2002: 47). Indeed, the *pro-growth* approach and entrepreneurial mentality are so entrenched in and imprinted upon the cultural and political landscapes of America that even the outwardly grassroots/collectivist policies used in shrinking cities to arrest decline are often rolled out in search of economic growth. This latter issue is engaged with as part of the next section, which discusses conventional policy responses to urban shrinkage in the United States. The common criticism of all of the policies examined herein is that their focus on economic growth neglects the socially and spatially uneven outcomes that necessarily accompany market-based development.

Pro-growth policies in U.S. shrinking cities

This section describes a non-exhaustive selection of conventional growth/decline management strategies for shrinking cities, through the lens of the American model of urban development. Several real-world examples are used to concretize the generic, "standardized" policy tools (Roth and Cunningham-Sabot 2012). Note well, though, that the following categories are not mutually exclusive – in fact, most of the examples could fall under more than one of the subsequent headings.

Large-scale urban development projects

Within the American model of urban development, municipalities are positioned against one another in what is essentially zero-sum competition for private investment. To bolster the relative competitiveness of their localities in these head-to-head transactions, one strategy that public authorities routinely roll out is the large-scale, subsidized, typically downtown development project (Swyngedouw, Moulaert, and Rodriguez 2002). While the specifics vary from case to case (see Table 6.1), these large-scale projects fit a general pattern and follow a general logic.

Table 6.1 Recent large-scale urban development projects in the United States

City	*Project – Type*	*Year*	*Approximate Cost*	*Cost per Person*[††] *(2010 population)*
Hartford, CT	*Hartford Stadium* – new downtown minor league baseball stadium	2015**	$56,000,000	$448.81
Detroit, MI	*Cobo Center* – major expansion to downtown convention center	2012[†]	$279,000,000	$390.88
Johnstown, PA	*Peoples Natural Gas Park* – indoor/outdoor concert and festival venue	2011	$4,000,000	$190.68
Roanoke, VA*	*Taubman Museum of Art* – downtown fine art museum	2008	$68,000,000	$700.80

* See note 1 regarding the inclusion of the Taubman Museum in this table. ** Denotes in progress [†] Denotes year renovated and/or expanded, not year built [††] This column reflects the total cost of the project, not of the public investments.

Namely, highly visible, "signature" developments such as waterfront improvements, downtown convention centers, sports stadia, business parks, and festivals are undertaken to produce a sense of spectacle, enchantment, and/or re-vibrancy in or proximate to areas experiencing decline. These endeavors are rationalized on at least two interrelated bases that embody neoliberal faith in competitive markets: (1) the Bilbao Effect and (2) trickle-down economics.

The **Bilbao Effect** refers to the successful economic transformation of the city of Bilbao (Biscay, Spain), which is often attributed to construction of the Bilbao Guggenheim Museum in the 1990s. In brief, Bilbao made its early mark as a shipbuilding, industrial, port city on the Spanish coast, and it grew rapidly throughout most of its history. Negatively affected by deindustrialization, however, the city lost over 14 percent of its residents between 1980 and 1990. The city's waterfront was hit especially hard, as abandoned industrial activities left the port area visibly scarred and untended. In response, and as part of a larger master-planning effort, the city hired renowned architect Frank Gehry to design an avant-garde waterfront building that would house a contemporary and modern art museum. Construction of Gehry's building was funded by city, state, national, and supranational public entities, while a private foundation agreed to manage the institution and provide for its operating expenses. The museum, which exhibits a contextually-sensitive design that fits well on the former industrial waterfront, was an instant success and has been a continued source of tourism and economic activity for the city of Bilbao since it opened in 1997 (Knox 2011).

Trickle-down economics refers to the more general notion that singular, large-scale developments like the Bilbao Guggenheim will improve the whole of an area through positive spillover effects and by acting as magnets for new economic opportunities and private investment. More colloquially, it adopts the premise that a "rising tide lifts all boats" (Teaford 2000). In the present context, large-scale development projects are believed to raise nearby property values, increase an area's aesthetic appeal, and thus stimulate reinvestment in a city. In this vein, it is claimed that large-scale projects kick-start a chain reaction of local economic development in the cities where they are implemented.

Although these rationalizations sound reasonable, in reality they tend to be repositories for misplaced confidence. For one, observers note that the Bilbao Effect would be better named the "Bilbao Anomaly", as numerous copycats of the design-led strategy have been unsuccessful at recreating the intended effect – even in cases where the same architect/designer (Frank Gehry) participated in the process (Rybczynski 2008). This result is plausibly because the Bilbao Guggenheim was part of a larger-scope, geographically inclusive master plan, whereas many of its offspring efforts appear to be single-shot, site-specific solutions (Swyngedouw, Moulaert, and Rodriguez 2002). In the second place, the trickle-down economics of large-scale developments rings of **design determinism**, which refers to a belief that development projects will alter individual behavior, despite not altering the contextual circumstances that generated pre-project conditions (Knox 2011). In other words, such projects do not outwardly improve the lives of most local residents, particularly the socio-economically disadvantaged, in ways that would substantively enhance their prospects for geographic and/or socio-economic mobility.

Nonetheless, despite criticisms of both the Bilbao and trickle-down rationales for "signature" economic development projects, the strategy remains quite popular and oft-practiced in growing and shrinking cities alike. This outcome is not altogether unexpected, as the approach fits comfortably in the American (neoliberal) model of development: it is premised on place-competition and attracting external investment to enhance a municipality's relative position in the spatial marketplace. As a consequence, signature projects are found all across the U.S. metropolitan landscape and have featured prominently in critical urban scholarship (del Cerro Santamaría 2013). Table 6.1 summarizes a handful of selected recent examples from U.S. shrinking cities.[2]

Tax foreclosure and public auctions

One of the most prominent physical manifestations of urban shrinkage is an oversupply of housing. Built environments that were constructed to accommodate multiples of a city's current population do not scale down at the same rate as the shrinking population. Rather, buildings are durable structures that remain fixed in place after their occupants vacate them. Further, because many spaces in shrinking cities are characterized by weak real estate markets (Schilling and Mallach 2012), vacant buildings sited in the least competitive neighborhoods of declining cities rarely sell in private transactions. Hence, these properties are tendentially *abandoned* by their owners, which is to say that deed holders abrogate their ownership responsibilities but do not officially alienate their rights. Under these conditions, owners discontinue paying property taxes, abandoned buildings suffer physical deterioration through both natural (e.g., weatherization) and manmade (e.g., vandalism) processes, and such structures turn into eyesores that perpetuate negative external images of affected neighborhoods.

The antisocial norms that are communicated by these circumstances are likely to erode a community's internal *social capital*, thereby leading to wider-spread local property abandonment, and, eventually, *decline* of the neighborhood (Ch. 4). Moreover, because neighborhoods are spatially interconnected, such outcomes are capable of spilling over to nearby communities, thus leading to even larger footprints of decline. Accordingly, addressing vacancy and abandonment is a critical issue for urban policy in shrinking cities (Hollander et al. 2009).

Many options are available to deal with vacant properties. For instance, an upcoming subsection considers the conversion of vacant parcels into community gardens and related projects. Further, the next chapter explores the creation of *land banks* – temporary public holding strategies – in the context of rightsizing. Yet, while these and other alternatives (Schilling and Logan 2008) exist, two options have come to predominate in shrinking cities: (1) tax foreclosure auctions and (2) structural demolition, often at a large scale. The former of these strategies is considered here, and the latter is taken up in the next subsection. The particulars of both strategies vary from context to context, but in the main they take on general forms.

Concerning tax foreclosure auctions, properties associated with delinquent local property tax payments, outstanding user fees, and/or unpaid municipal water and sewer charges may be foreclosed upon by local governments. Foreclosure proceedings are typically initiated after years of property neglect and abandonment and only after several attempts are made to notify the legal deed holder of the debt and collect payment (Silverman, Yin, and Patterson 2013). Where these attempts are unsuccessful – a regular outcome in shrinking cities – foreclosure may ensue, and successfully foreclosed-on properties are deeded to the local public authority.

By and large, the next significant step in this process is public auction. Local governments inventory their holdings and attempt to sell off individual properties to private buyers in annual or semi-annual market exchanges. Such auctions have at least two objectives: first, to collect payments sufficient to cover each property's outstanding debts; and second, to return erstwhile vacant and abandoned structures to the municipal tax rolls by placing them in the hands of private investors. While in some cases prosocial neighborhood residents use these auctions to acquire adjoining lots and improve their local communities, the auction system regularly leads to land and property speculation, absentee ownership, and associated potentially antisocial outcomes (Lawson and Miller 2012: 39).

Further, the auctions rarely benefit the most distressed urban communities, as properties in these "weak market" or "dead zone" areas go unsold and continue to tear away at the social and physical fabrics of their neighborhoods (Knight and Weaver 2015). Notwithstanding these criticisms, shrinking cities regularly adopt the tax foreclosure auction strategy, which is rationalized on its [supposed] capacity to attract new private investment. Consistent with the supply and demand assumptions of neoclassical economics, private buyers are assumed to *want* to buy city-owned real estate at low, often negligible auction prices. These private transactions, it follows, will revalorize distressed urban spaces in a manner consistent with the trickle-down economics logic described earlier.

Whether such outcomes bear out in practice hinges on local contextual variables, particularly the extent to which auction winners positively influence social norms in their neighborhoods. For instance, cases alluded to earlier for which neighborhood residents acquire foreclosed properties at city auctions have been linked to stabilization or improvement (Lawson and Miller 2012). In these scenarios, residents presumably enforce local norms, improve visible conditions in their neighborhoods, and positively affect their communities' capacities for collective action (Weaver and Holtkamp 2015). However, because these transactions usually involve converting acquired parcels into gardens or larger yards, they contribute to de-densification and do not generate the level of public tax revenue of, say, another house (Hollander et al. 2009). As a consequence, public authorities show a preference for outcomes where open market auction competition allocates city-owned properties to developers (Lawson and Miller 2012). Yet, this preference for a pro-growth outcome fails to consider what happens to local social fabrics when absentee owners purchase properties in neighborhoods that are under pressure to change, or when speculators (or cities) allow properties to remain vacant as they wait for market conditions to improve (Hackworth 2014).

Examples of the public auction system can be found in almost any U.S. city. For present purposes, two recent studies offer useful illustrations. First, Dewar and Thomas (2012) examined the public auction process in Detroit, Michigan for a specific distressed neighborhood from 2002 through 2010. During that time, only 18 percent of properties offered at auction sold, and many of the buyers were real estate speculators. As such, the vast majority of vacant and abandoned parcels in the neighborhood remained just that. Discouraged by this outcome, local residents began (without title) to modify some of these properties in ways that suited their needs. Gardens were planted on several vacant parcels, and other lots were used (illegally) as dump sites. Notably, the bottom-up nature of these modifications made them relatively adaptive land uses in the neighborhood. While interviewing locals, Dewar and Thomas (2012) found that residents spoke somewhat acceptingly of the city's lack of involvement in revitalizing their neighborhood, for this meant that their (untitled) modifications to abandoned properties were not challenged. The implication is that, were the auction system to produce the desired pro-growth outcome, whereby properties are allocated to private developers, then the eventual land uses might not conform to these local norms. In other words, contrary to the *pro-growth* approach, which holds that economic growth is a *successful* policy outcome, pro-growth "solutions" might be less adapted to the neighborhood in this case study, given their relative inattention to local context.

Second, the city of Buffalo, New York, holds regular *in rem* auctions to sell off its foreclosed properties (Silverman, Yin, and Patterson 2012). In a spatial analysis of annual *in rem* sales data for six years of auctions, Knight and Weaver (2015) identified an apparent "dead zone" in the east-central part of the city. The authors used a spatial scan statistic to identify, for each year of auction data, geographic clusters of high and low sales activity. Viewing the results collectively, the authors found that 88 percent of *all* properties that were included in low sales (unsold property) clusters during any of the six auction years fell within the geographic area. That is, the "dead zone" represents a space where city-owned properties fail to attract private bidders at public auctions. This finding implies that at present the space has little or no market potential to draw in outside private investment. Not surprisingly, the affected neighborhoods are some of the most distressed and impoverished places in Buffalo (Silverman, Yin, and Patterson 2012; Frazier, Bagchi-Sen, and Knight 2013; Frazier and Bagchi-Sen 2015). In contrast, the bulk of high foreclosure sales clusters were detected to the west and north of the dead zone, in areas that are known to be stable or gentrifying (Ch. 4). The implication is that relying too heavily on the private market to stimulate growth in declining cities will continue to favor "profitable" spaces over those that are characterized by weak or no market demand.

Massive demolition programs

A "favored" vacancy management strategy of shrinking cities, which has been pursued with "great vigor" since the turn of the twenty-first century, is demolition (Ryan 2012). Targeting chronically vacant and abandoned structures, including city-owned foreclosures, "massive demolition programs" aim to clear land for private development, while also reducing a city's overall vacancy rate (Accordino and Johnson 2000). As with the other policies considered above, such programs appear to be well-intentioned and necessary efforts to improve conditions in shrinking cities – in this case by ridding their landscapes of nuisance properties. However, in the context of the American model of urban development, the large-scale demolition approach tends to operate in service to competition and public entrepreneurship. For example, the well-publicized "5 in 5 Demolition Plan" in Buffalo, New York, set out to demolish 5,000 vacant properties in the course of five years, beginning in late 2007. This seemingly arbitrary quantitative target was set not for context-specific reasons (Schilling and Logan 2008) but so the city's overall vacancy rate would creep down to match those of other upstate New York cities (Brown 2007). Stated another way, the program sought to increase the city's competitive position relative to its peer group. Moreover, like comparable programs in Philadelphia (Ryan 2012) and Detroit (Herscher 2012), the Buffalo strategy was largely speculative – the city demolished structures as it was able, not for evidence-based reasons. As Ryan (2012: 182) observes:

> Apart from being very costly . . . the [large-scale] demolition strategies carried out in shrinking cities after 2000 suffered from a basic flaw: they were driven by a simple imperative to demolish vacant buildings, with little idea about what the vacant lots would be used for . . . cities like Philadelphia and Buffalo simply demolished buildings where it was politically expedient or where life safety issues drove demolition crews to act quickly. This parcel-by-parcel removal strategy effectively led to random vacant lots scattered among remaining properties . . . [O]nce demolished, vacant parcels were more or less unmarketable, both because they were scattered and because they were in depressed neighborhoods where there was little market demand for land either before or after demolition. Demolition removed abandoned structures, but it did not generate spatial strategies for depressed neighborhoods, and it did not create development markets where there had been none before.

What ought to stand out thus far is that three prominent, but distinct, policy strategies in shrinking cities – signature development projects, foreclosure auctions, and demolition programs – are all critiqued along very similar lines. Namely, in their "American model" forms, they are acontextual and place much confidence in the ability of the market to catalyze growth in distressed neighborhoods. Building on the latter of these, they also put much faith in the belief that private developers will respond to investment incentives (low auction prices, cheap vacant land, etc.) that are created in depressed neighborhoods that otherwise have little or no market appeal. The final category of policies considered here – art spaces and community gardens – involves a much more bottom-up and potentially context-sensitive approach. With this strategy, reinvestment *does* occur in distressed or weak market neighborhoods. On its face, then, the approach looks altogether different from the former three. Interestingly, however, it has been subject to similar lines of criticism – or, more accurately, skepticism – suggesting that it might be necessary to work outside the American model, in more cooperative governance frameworks, to manage change in shrinking cities (Chs. 7–8).

Community-based initiatives: art spaces, pocket parks, and community gardens

Foreclosure sales and structural demolitions, while popular with policymakers, are not the only means for dealing with vacancy and abandonment. Increasing attention is being paid to grassroots efforts to reclaim and reuse vacant property, including art projects (Herscher 2012), pocket parks (Foo et al. 2014), and community gardens (Lawson and Miller 2012). This section discusses such bottom-up actions using the generic term *community-based initiatives*. Lumping these and like approaches together is done for parsimony and is not meant to downplay the diversity of grassroots initiatives that exist in urban environments. The explicit naming of three strategies here is done both for illustrative purposes, and because they coincide with examples presented in this subsection. Moreover, these specific approaches, despite their distinctions, often overlap. For example, a 2010 community-based project in Buffalo, New York, involved (1) clearing the land on two adjoining vacant lots in a distressed neighborhood; (2) replacing what was on those lots with native vegetation, benches, and a stone walking path; and (3) working with the owner of an abutting building to commission a mural on a wall that faced the lots. In this way, the community initiative created a pocket park *and* an art space, though the stakeholders referred to the outcome as a "community garden" (City of Buffalo Common Council 2010). Thus, it had elements of all three types of actions enumerated above. The point is that, in many cases, it is difficult to discretely classify community-based initiatives into a single category.

On that foundation, the common thread that does seem to run through most community-based initiatives is bottom-up action. Initiatives are almost always conceived of and driven by citizens or organizations from within a targeted neighborhood (Lawson and Miller 2012). Observations made by Dewar and Thomas in Detroit (2012; see above) suggest that many of these efforts begin as "illegal" or untitled uses of abandoned property. Citizens who become frustrated with chronically vacant parcels, especially their dilapidated conditions, occasionally "take back" problem properties before gaining legal authority to do so. An example of this sort of reclamation is **guerrilla gardening,** whereby residents come together to beautify vacant lots through landscaping and other means without [immediate] concern for who owns the land (Reynolds 2008). This prosocial deed most often originates inside a transitioning neighborhood, and it therefore sends a powerful, collective-minded signal about internal social capital to other neighborhood residents. It is consequently of no surprise that researchers have found guerrilla gardening to be capable of improving conditions and internal relations in affected neighborhoods, especially along the dimensions of social justice (Milbourne 2012).

Importantly, then, the organic and emergent qualities of community-based initiatives imply that they are more likely to adapt to their local contexts than strategies imposed from the top-down (Marshall 2009). Further, they represent capital inflows into what might otherwise be "dead zones" for market reinvestment. For these reasons, they are attractive strategies for urban regeneration. Nonetheless, given their embeddedness in the American model of urban development, there is skepticism about both the long-term viability of such projects, and whether they are able to scale up beyond the level of a neighborhood. With respect to long-term viability, there are at least two areas of concern. First, community-based initiatives generally require a minimum stock of internal social capital to succeed; and second, like economic capital, social capital exhibits an uneven geographic distribution (Ch. 4). Hence, following from prior critiques of the public sector's overdependence on private capital to revitalize urban areas, relying exclusively on social capital to manage neighborhood change invariably can leave some marginal places behind. In addition, even where outside intervention is used to jumpstart community-based initiatives in marginal spaces, such efforts will only be sustainable if residents buy into them.

Consider the Buffalo example from the beginning of this subsection. In that case, a public-private-nonprofit partnership gained access to adjoining city-owned vacant lots to erect a community garden (Fig. 6.1). The project site fell within a wider geographic territory whose short- and long-term vitality was of direct interest to all partners. That being said, the success of a community garden hinges on

Figure 6.1 Volunteers create a community garden on adjoining vacant lots in Buffalo, New York (2011; photo courtesy of CityCorps Buffalo).

Figure 6.2 The same community garden in 2014. The current state is a significant improvement over pre-project abandonment conditions; but the site no longer has the appearance of an actively maintained pocket park or community garden (source: Google Street View).

resident participation and collective action (Lawson and Miller 2012). This stipulation matters to the extent that residents of the block on which the garden was constructed ultimately did not participate in the project in a substantive way. While there were efforts to engage residents in the process, including the eventual management of the site, the work was carried out by volunteers who were external to the neighborhood. Through donations, grants, and start-up support from the local government and a local business owner, the volunteer team successfully managed the garden for well over a year. However, when funding dried up and stakeholders stepped down from their positions, the garden was no longer being actively maintained at the desired level. At present, what remains are a few surviving trees and a fading mural, while the benches, stones, and a preponderance of the planted vegetation have all disappeared (Fig. 6.2).

A second area of concern regarding the long-term viability of community-based initiatives is their temporality vis-à-vis changing market conditions. Explicitly, within the fragmented, competitive, and entrepreneurial American metropolis, the mantra of local government is that of "highest and best use" (Blomley 2004). When vacant and abandoned property attracts private investment, community-based land uses such as pocket parks and neighborhood gardens tendentially lose out to the prospect of taxable improvements (Lawson and Miller 2012). As a result, community-based initiatives are regularly perceived [by public authorities] as temporary means for upholding property values during times of uncertainty and downward market pressure. Their role is to "create a perception of stability and . . . increase the market value of vacant properties for potential investors" (Hollander et al. 2009: 227).

The observation that community-based projects are perhaps placeholders for economic development syncs with the second area of skepticism alluded to above, regarding the ability of these efforts to scale up beyond the level of a single neighborhood. Specifically, empirical evidence showing that amenities like pocket parks and art spaces stabilize or even increase nearby property values (Conway et al. 2010) has strengthened interest in "culture-led" urban regeneration policies. Whereas culture-led regeneration is sometimes discussed in terms of large-scale urban development projects (see above) aimed at attracting members of the "creative class" (Roth and Cunningham-Sabot 2012), the same line of argument applies at smaller scales. In both cases, strategies are geared toward gentrifying inner-city neighborhoods – that is, drawing relatively higher income and/or better educated individuals into distressed urban spaces. The interest in such outcomes is apparent: gentrification processes create demand in weak markets; new demand raises property values; higher property values generate more tax revenue for local governments; and, thus, the relative positions of gentrifying cities are improved in the competitive, pro-growth urban order.[3]

In this context, amenities that outwardly begin as community-based initiatives – e.g., art spaces, community gardens, and pocket parks – undergo *commodification* and are reframed as tools of economic growth and revitalization (Hollander et al. 2009; Herscher 2012). As an urban development strategy, cultural capital is used as bait for economic capital. Despite the bottom-up, collective origins of the various approaches presented in this subsection, in practice, political attempts to scale such efforts up to the city level are regularly characterized by overt or latent market fundamentalism (Herscher 2012). This adherence to the American model of urban development frequently obscures the negative side of growth [via gentrification] from view. In particular, the very social conditions that produce sought-after (commodified) cultural amenities are occasionally uprooted during gentrification. Existing households in affected neighborhoods sometimes lack the financial means to keep up with escalating property taxes and the rising maintenance standards that can accompany gentrification (Newman and Wyly 2006). As a result, while culture-led urban revitalization arguably improves many aspects of affected neighborhoods, the pro-growth, free market environment in which such improvements take place can give way to heightened, plausibly more contentious, forms of social and economic polarization (Kantor 2010).

Community Development Block Grant expenditures: evidence of the pro-growth approach in shrinking cities?

As part of the analysis of population shrinkage in Chapter 2, we identified all *census places* (Table 2.1) in the conterminous United States that reported 50,000 or more residents in the most recent (2010) decennial census. For all intents and purposes, this set of places (n = 707) represents *cities* in the conterminous United States whose populations make them eligible to receive funding from various federal government urban initiatives (Box 6.1 and Ch. 2). By joining four decades (1970–2010) of census tract-level data from the Brown University LTDB (see Logan, Xu, and Stults 2014) to the current boundaries of these census places, we identified thirty-one cities that have endured *persistent* and *severe* population shrinkage since at least 1970 (Fig. 2.5).

One U.S. federal urban initiative that provides streams of funding to principal cities and other places with 50,000 or more residents is the **Community Development Block Grant** (CDBG) Entitlement Program. The CDBG was established by federal legislation in 1974 and is administered by the U.S. Department of Housing and Urban Development (HUD). Presently, HUD grants CDBG funds to eligible municipalities for community development activities including, but not limited to, economic development, public improvements, and public services. Box 6.1 reproduces descriptions of CBDG program eligibility requirements and permitted activities from online HUD resources. Importantly, observe that CDBG-funded activities must attend to one or more national HUD objectives, such as benefiting low- and moderate-income persons, preventing "blight", and neutralizing immediate threats to a community's "health and wellness" (Box 6.1).

Due to the eligibility requirements and formula used to distribute CDBG funds, different cities receive different amounts of (or zero) funding in a given grant period. Moreover, once HUD releases CDBG funds to the local government of a grantee city, local decision-makers for that city then have considerable discretion over how they allocate the funds across the eligible categories of activities (Box 6.1). Hence, expenditures by authorized activity will vary from place to place. Consequently, it may be possible to investigate city-level patterns of CDBG-related expenditures to determine whether, for example, *shrinking cities* are more likely than non-shrinking cities to allocate these public resources (CDBG funds) to economic development activities. Recall that within the above context, non-shrinking places are assumed to be more "successful" than shrinking places at attracting private economic investment. Therefore, non-shrinking cities are presumably less reliant on government to stimulate economic growth. By comparison, because shrinking cities are perceived to be "unsuccessful" at attracting economic investment – and in the context of the *pro-growth* approach that underlies the *American model of urban development* – decision-makers in these cities are perhaps more driven to use public dollars to catalyze growth. In other words, the pro-growth approach suggests that decision-makers in shrinking cities ought to have a greater incentive than their counterparts in non-shrinking cities to leverage all available means to generate growth.

Following a similar line of reasoning, scholars in the shrinking cities research community observe that pro-growth approaches in urban policy arenas cause decision-makers to overlook the people who remain in shrinking places (Ryan 2012). Shrinking places are never fully emptied. By choice or because of mobility constraints, scores of people continue to live in places that endure shrinkage and/or decline. As a result, political preoccupations with attracting new economic growth, typically from external sources, tend to neglect the human, social, cultural, built, and natural assets that are already present *within* these communities. On that note, some places that have successfully pushed back against processes of *decline* (Ch. 3) have done so by giving preference to improving the *quality* of their public services and infrastructure, over increasing the *quantity* of their economic activity (Ryan 2012). For these reasons, in addition to our expectation that shrinking cities are more likely than their counterparts to use CDBG funding for economic development, we further hypothesize that shrinking places allocate smaller percentages of CBDG dollars to public service and infrastructure improvements relative to non-shrinking places.

To evaluate these hypotheses, we downloaded the most recent version of the geospatial "HUD Grantee Activities" database, which is available to the public through the HUD website.[4] This database reports on all CDBG-funded activities (including amount of funding allocated), broken down by permitted activity type (Box 6.1), and by census tract, from 1990 through the end of 2013. By applying the techniques that were described in Chapter 2 to join census tract data to city boundaries – where cities are taken to be census places with 50,000 or more residents in 2010 – we are able to aggregate all tract-level CDBG data from the HUD database to the city-level and join it to our census place-level database from Chapter 2. The result of this process is a set of 680 cities for which CDBG-funded activities are reported in the HUD Grantee Activities database. These 680 cities are mapped in Figure 6.3, where they are symbolized according to the nature of the population change they experienced between 1970 and 2010. In all, there are: (1) 31 *shrinking cities* whose populations decreased by 25 percent or more; (2) 101

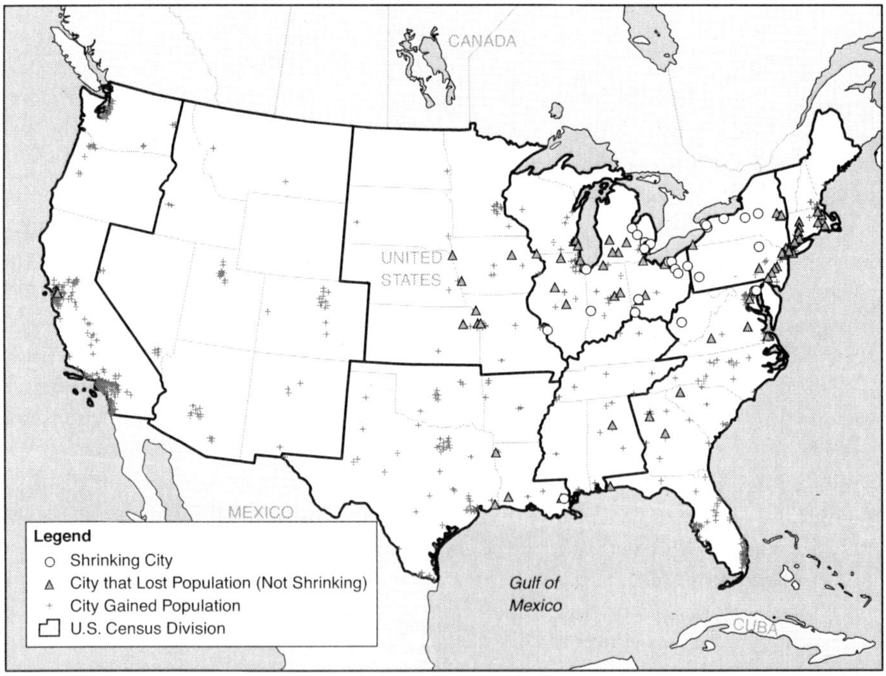

Figure 6.3 HUD CDBG grantee cities (n = 680), by nature of population change (1970–2010)

cities that *lost population* but did not meet our adopted threshold to qualify as shrinking (Ch. 2);[5] and (3) 548 cities that *gained population* over the four-decade period from 1970 to 2010.

The point of grouping places according to whether they *shrank*, *lost population* but did not experience shrinkage, or *gained population* for this analysis is to trace out the full implications of the conceptualization behind policy making described in the above paragraphs. Namely, if gaining population is tantamount to "success" in the American model of development, then losing population – in any form – is "failure" (Leo and Anderson 2006). Thus, in the context of our aforementioned hypotheses, we should expect both *shrinking cities* and cities that *lost population* (but did not endure severe shrinkage) to display different CDBG spending patterns relative to cities that *gained population*. Furthermore, because *severe* and *persistent* shrinkage is presumably more of a "failure" than less substantial population loss, then expenditure patterns should also differ between shrinking cities and these comparatively marginally depopulating cities. However, expenditures in these two types of settlements should still be more similar to one another than they are to places that gained population.

To test these hypotheses and implications for each city depicted in Figure 6.3, we created two variables: (1) the percentage of the total CDBG funding reported for the city in the HUD database that was allocated to economic development and (2) the percentage of total CDBG funding reported for the city in the HUD database that was allocated to public improvements or public services. Because these variables exhibit skewed distributions within the groups of cities, we rely on the nonparametric Kruskal-Wallis (K-W) test to facilitate the analyses. The K-W test is a nonparametric analogue to one-way analysis of variance (ANOVA), such that it can loosely be thought of as a tool for comparing the median of a variable between groups.

Table 6.2 presents the results of the K-W tests for each of the two variables enumerated in the preceding paragraph. In both cases, the p-values are small enough

Table 6.2 Kruskal-Wallis (K-W) tests for equality of medians, by place, by nature of the place's population change from 1970–2010

	Shrank	*Lost Population*	*Gained Population*	*p-value*
Economic development expenditures (% of total CDBG funds)	**5.9**	4.3	0.4	<0.001
Public service and public improvement expenditures (% of total CDBG funds)	30.8	33.3	**41.8**	0.002
n	31	101	548	

Notes: The values reported in the table are median percentages (e.g., 5.9 in the first cell refers to 5.9 percent). Post hoc pairwise tests were carried out with Conover's test for multiple comparisons of independent samples (Pohlert 2014) using a false discovery rate p-value correction method (Benjamini and Hochberg 1995). **Bold text** indicates the largest value in each row (i.e., the largest group median for the given variable). Universe: cities with 50,000 or more residents, for which CDBG funds were reported in the HUD Grantee Activities geospatial database to reject the null hypotheses of equal group medians at all conventional levels of statistical significance. Thus, we can confidently conclude that the three different types of cities make different median allocations of CDBG dollars to economic development activities and public service and infrastructure improvements. Crucially, these significant differences in CDBG expenditure patterns are all in the hypothesized directions. Of the three classes of places, shrinking cities allocate the largest median percentage of CDBG funds to economic development and the smallest median percentage of CDBG funds to public service and infrastructure improvements.

Recall from Chapter 3, though, that the K-W test is an omnibus test. Thus, while the results in Table 6.2 allow us to conclude that these median CDBG allocations in shrinking, marginally depopulating, and growing cities are different, they do not allow us to say specifically that one type of city (e.g., shrinking) has a different median allocation from a second type of city (e.g., city that gained population). Accordingly, as we did in Chapter 3, post hoc analyses were performed on the K-W tests from Table 6.2 with Conover's test for multiple comparisons (Pohlert 2014) and a false discovery rate p-value correction technique (Benjamini and Hochberg 1995). The p-values from these post hoc tests are reported in Table 6.3. These results are highly suggestive of a *pro-growth* approach in median CDBG allocations wherein population loss is perceived as "failure" to attract economic investment. To be sure, the results imply that differences in the *severity* of population loss are not associated with significant differences in median CDBG allocations. When it comes to CDBG expenditures, any population loss – severe or otherwise – appears to be interpreted as "failure": both shrinking cities and marginally depopulating cities allocate significantly more median CDBG dollars to economic development than growing cities but in ways that are not significantly different from one another (note the large p-values from the post hoc comparisons of shrinking cities and cities that lost population but did not shrink). The opposite pattern of results holds for expenditures on public service and infrastructure improvements: shrinking and marginally depopulating cities allocate significantly smaller median percentages of their CDBG funds to these activities relative to growing cities[6] but again in ways that are not significantly distinguishable from one another. While the evidence from Tables 6.2 and 6.3 should be considered circumstantial and preliminary, the patterns revealed in the city-level CDBG spending data readily

Table 6.3 p-values for pairwise post hoc tests from Table 6.2

	Lost Population	*Gained Population*
Economic development expenditures		
Shrank	0.743	0.001**
Lost Population	—	<0.001***
Public service and public improvement expenditures		
Shrank	0.910	0.109†
Lost Population	—	0.003**

***$p < 0.001$ **$p < 0.010$ †$0.100 < p < 0.110$

Box 6.1 Community Development Block Grant (CDBG) eligibility requirements and permitted activities

"Eligible grantees are as follows:

-Principal cities of Metropolitan Statistical Areas (MSAs)
-Other metropolitan cities with populations of at least 50,000 . . .

Eligibility for participation as an entitlement community is based on population data provided by the U.S. Census Bureau and metropolitan area delineations published by the Office of Management and Budget. The U.S. Department of Housing and Urban Development (HUD) determines the amount of each entitlement grantee's annual funding allocation by a statutory dual formula, which uses several objective measures of community needs, including the extent of poverty . . . and population growth lag in relationship to other metropolitan areas."

Source: https://www.hudexchange.info/programs/cdbg-entitlement/cdbg-entitlement-program-eligibility-requirements/

CDBG funding may be allocated to the following activities, which are all documented in the "HUD Grantee Activities" geospatial database mentioned in the text:

"**Acquisition Activity** – CDBG activity related to acquisition, including disposition, clearance and demolition, and clean-up of contaminated sites/brownfields.

Economic Development Activity – CDBG activity related to economic development, including commercial or industrial rehab, commercial or industrial land acquisition, commercial or industrial construction, commercial or industrial infrastructure development, direct assistance to businesses, and micro-enterprise assistance.

Housing Activity – CDBG activity related to housing, including multi-family rehab, housing services, code enforcement, operation and repair of foreclosed property and public housing modernization.

Public Improvements Activity – CDBG activity related to public improvements, including senior centers, youth centers, parks, street improvements, water/sewer improvements, child care centers, fire stations, health centers, non-residential historic preservation, etc.

Public Services Activity – CDBG activity related to public services, including senior services, legal services, youth services, employment training, health services, homebuyer counseling, food banks, etc.

Other Activity – CDBG activity related to urban renewal completion, non-profit organization capacity building, and assistance to institutions of higher education."

Source: https://www.huduser.gov/portal/datasets/gis/granteeact.html

support the argument that decision-makers in shrinking cities possess a pro-growth approach that promotes economic growth over most other urban or regional policy objectives (Leo and Anderson 2006; Kantor 2010; Ryan 2012).

Concluding remarks

In this chapter, we examined a variety of urban policy responses to shrinkage and decline within the context of the current ideology guiding policymaking in the United States. The selected policy instruments by no means constitute the totality of a city's growth (decline) management toolkit. However, all of the policies are commonly accepted by leaders in shrinking cities as tools for stymying shrinkage and decline and reversing the negative effects of these phenomena. Yet, when situated within the neoliberal American model of urban development, each policy instrument – even the comparatively organic and bottom-up community-based initiatives – has also served as a strategic lever for generating economic growth via market transactions.

One of the primary critiques of the neoliberal belief that the market optimally allocates scarce resources is that, in the pursuit of accumulation (growth), disadvantaged persons and populations are necessarily left behind (Harvey 1989). That is, markets are not as impartial and universally accessible as they are often portrayed to be. Rather, while under certain strict assumptions neoclassical economic theory demonstrates that markets allocate resources more *efficiently* than any other mechanism (i.e., they ensure that all mutually beneficial transactions take place), they rarely if ever allocate those resources *equitably* among members or segments of society. The growth-oriented goal of **efficient allocation** circumvents the important social issue of distributional equity (Daly and Farley 2004). Additionally, note that the U.S. model of urban change views shrinkage and decline as *problems* in need of *solutions* (Kantor 2010). Shrinkage and decline are stigmas. They are unenviable statuses – the opposite of growth; the opposite of "winning" (Leo and Anderson 2006). This perception – that cities must be growing to be desirable places in which to live – forms the bedrock, and covers the full extent, of the American cultural and political landscapes. In this way, the pro-growth approach is institutionalized in the philosophy and management strategies of shrinking cities (Schatz 2012). This observation implies that moving beyond the American model may require a paradigm shift. Planning scholars and practitioners have been making significant progress in bringing an alternative approach to the forefront of the shrinking cities discourse for at least a decade. Framing their work with headings such as "rightsizing", "greening", "smart decline", "sustainable urbanism", "urban regeneration", and "new renewal", these scholars are attempting to move beyond planning's pro-growth approach to envision new, more compact, equitable, and civil urban societies for America's shrinking cities. The next chapter provides an overview of these ideas, along with a look at how some of their recommendations have shaped recent urban planning and development efforts in selected shrinking cities.

Notes

1 *Actually existing neoliberalism* in contemporary society is a mixture of roll-back and roll out initiatives, but such that it departs from its "pure" form in which government is viewed as a hindrance to economic development (Brenner and Theodore 2002).

2 Roanoke, Virginia, does not meet the criteria/threshold method used in Chapter 2 to identify *shrinking* places. The city reached its peak population of 100,220 persons in 1980, and the 2010 decennial census reported that 96,919 persons were living in Roanoke at that time. Thus, while the city has lost population over the past thirty years, the loss has not been *severe*. The reason for including this relatively stable-population city in Table 6.1, then, deals with the urban development project that is highlighted. The Taubman Museum represents one of the closest attempts to copy-and-paste the Bilbao Guggenheim Museum into a U.S. city. Roanoke even retained a protégé of Frank Gehry, Randall Stout, to lead the design efforts (see Knox 2011 for more information).

3 This line of reasoning, while necessarily simplified, is emblematic of the thinking that underlies efforts to repopulate shrinking cities with, above all else, members of the so-called "creative class" (Sager 2011).

4 See: http://www.huduser.gov/portal/datasets/gis/granteeact.html

5 In other words, these 101 cities shrank according to the *binary method* of identifying shrinkage, but not the *threshold method*.

6 NB: this conclusion requires a caveat, insofar as the post hoc analyses reveal the difference in medians between shrinking and growing cities is only significant at an 89 percent level of confidence. In general, a 90 percent level of confidence is considered to be the lowest conventional level of confidence to claim statistical significance.

References

Accordino, John, and Gary T. Johnson. 2000. "Addressing the Vacant and Abandoned Property Problem." *Journal of Urban Affairs* 22 (3):301–315.

Beauregard, Robert A. 2006. *When America Became Suburban*. Minneapolis: University of Minnesota Press.

Benjamini, Yoav, and Yosef Hochberg. 1995. "Controlling the False Discovery Rate: A Practical and Powerful Approach to Multiple Testing." *Journal of the Royal Statistical Society. Series B (Methodological)* 57 (1):289–300.

Blomley, Nicholas K. 2004. *Unsettling the City: Urban Land and the Politics of Property*. New York: Routledge.

Brenner, Neil, and Nik Theodore, eds. 2002. *Spaces of Neoliberalism: Urban Restructuring in Western Europe and North America*. Maiden, MA: Blackwell Publishers Ltd.

Brown, Mayor Byron. 2007. *Mayor Brown's "5 in 5" Demolition Plan*. Buffalo, New York. https://www.ci.buffalo.ny.us/files/1_2_1/Mayor/PublicPolicyPublication/5in5_DemoPlan.pdf

Conway, Delores, Christina Q. Li, Jennifer Wolch, Christopher Kahle, and Michael Jerrett. 2010. "A Spatial Autocorrelation Approach for Examining the Effects of Urban Greenspace on Residential Property Values." *The Journal of Real Estate Finance and Economics* 41 (2):150–169.

Daly, Herman E., and Joshua Farley. 2004. *Ecological Economics: Principles and Applications*. Washington, DC: Island Press.

del Cerro Santamaría, Gerardo. 2013. *Urban Megaprojects: A Worldwide View*. Bingley: Emerald Group Publishing Limited.

Dewar, Margaret, Christina Kelly, and Hunter Morrison. 2012. "Planning for Better, Smaller Places after Population Loss: Lessons from Youngstown and Flint." In *The City after Abandonment, University of Pennsylvania Press, Philadelphia, PA*, edited by Margaret Dewar and June Manning Thomas, 290–316. Philadelphia: University of Pennsylvania Press.

Dewar, Margaret, and June Manning Thomas, eds. 2012. *The City after Abandonment.* Philadelphia: University of Pennsylvania Press.

Emery, Mary, and Cornelia Flora. 2006. "Spiraling-Up: Mapping Community Transformation with Community Capitals Framework." *Community Development* 37 (1):19–35.

Foo, Katherine, Deborah Martin, Clara Wool, and Colin Polsky. 2014. "Reprint of 'the Production of Urban Vacant Land: Relational Placemaking in Boston, Ma Neighborhoods'." *Cities* 40 (Part B):175–182.

Frazier, Amy E., and Sharmistha Bagchi-Sen. 2015. "Developing Open Space Networks in Shrinking Cities." *Applied Geography* 59:1–9.

Frazier, Amy E., Sharmistha Bagchi-Sen, and Jason Knight. 2013. "The Spatio-Temporal Impacts of Demolition Land Use Policy and Crime in a Shrinking City." *Applied Geography* 41:55–64.

Glassman, J. F. 2009. "Structural Power, Agency, and National Liberation." In *Geographic Thought: A Praxis Perspective*, edited by Marvin Waterstone and George Henderson, 308. New York: Routledge.

Großmann, Katrin, Marco Bontje, Annegret Haase, and Vlad Mykhnenko. 2013. "Shrinking Cities: Notes for the Further Research Agenda." *Cities* 35:221–225.

Hackworth, Jason. 2014. "The Limits to Market-Based Strategies for Addressing Land Abandonment in Shrinking American Cities." *Progress in Planning* 90:1–37.

Harvey, David. 1989. "From Managerialism to Entrepreneurialism: The Transformation in Urban Governance in Late Capitalism." *Geografiska Annaler. Series B. Human Geography* 71 (1):14.

Herscher, Andrew. 2012. *The Unreal Estate Guide to Detroit.* Ann Arbor: University of Michigan Press.

Hollander, Justin B., Karina Pallagst, Terry Schwarz, and Frank J. Popper. 2009. "Planning Shrinking Cities." *Progress in Planning* 72 (4):223–232.

Kantor, Paul. 1995. *The Dependent City Revisited: The Political Economy of Urban Development and Social Policy.* 2nd ed. Boulder: Westview Press.

Kantor, Paul. 2010. "City Futures: Politics, Economic Crisis, and the American Model of Urban Development." *Urban Research & Practice* 3 (1):11.

Kelo, V. City of New London 545 U.S. 469 [2005].

Knight, Jason, and Russell Weaver. 2015. "Using Tax Foreclosure Auction Results to Detect 'Dead Zones' in Shrinking Cities." Annual Meeting of the Association of American Geographers, Chicago.

Knox, Paul L. 2011. *Cities and Design, Critical Introductions to Urbanism and the City.* New York, NY: Routledge.

Lawson, Laura, and Abbilyn Miller. 2012. "Community Gardens and Urban Agriculture as Antithesis to Abandonment – Exploring a Citizenship-Land Model." In *The City after Abandonment*, edited by Margaret Dewar and June Manning Thomas, 400. Philadelphia: University of Pennsylvania Press.

Leo, Christopher, and Kathryn Anderson. 2006. "Being Realistic about Urban Growth." *Journal of Urban Affairs* 28 (2):169–189.

Logan, John, and Harvey Molotch. 1987. *Urban Fortunes: The Political Economy of Place*. Berkeley: University of California Press.

MacLeod, Gordon. 2002. "From Urban Entrepreneurialism to a 'Revanchist City'? On the Spatial Injustices of Glasgow's Renaissance." *Antipode* 34 (3):602–624.

Madanipour, Ali. 2006. "Roles and Challenges of Urban Design." *Journal of Urban Design* 11 (2):173–193.

Marshall, Stephen. 2009. *Cities Design & Evolution*. London, UK: Routledge.

Milbourne, Paul. 2012. "Everyday (in) Justices and Ordinary Environmentalisms: Community Gardening in Disadvantaged Urban Neighbourhoods." *Local Environment* 17 (9):943–957.

Newman, Kathe, and Elvin K. Wyly. 2006. "The Right to Stay Put, Revisited: Gentrification and Resistance to Displacement in New York City." *Urban Studies* 43 (1):23–57.

Peck, Jamie, Nik Theodore, and Neil Brenner. 2009. "Neoliberal Urbanism: Models, Moments, Mutations." *SAIS Review of International Affairs* 29 (1):49–66.

Peck, Jamie, and Adam Tickell. 2002. "Neoliberalizing Space." *Antipode* 34 (3):380–404.

Pohlert, Thorsten. 2014. The Pairwise Multiple Comparison of Mean Ranks Package (PMCMR). https://cran.r-project.org/web/packages/PMCMR/vignettes/PMCMR.pdf

Roth, Hélène, and Emmanuèle Cunningham-Sabot. 2012. "Shrinking Cities in the Growth Paradigm: Towards Standardised Regrowing Strategies?" In *Parallel Patterns of Shrinking Cities and Urban Growth Spatial Planning for Sustainable Development of City Regions and Rural Areas*, edited by Robin Ganser and Rocky Piro, 59–72. Burlington: Ashgate Publishing Ltd.

Ryan, Brent D. 2012. *Design after Decline: How America Rebuilds Shrinking Cities*. Philadelphia: University of Pennsylvania Press.

Rybczynski, Witold. 2008. "Architectural Branding." *Appraisal Journal* 76 (3):279–284.

Schatz, Roland. 2012. "Overcoming the Risk of Stereotypes: How Strategic Communications Can Facilitate Sustainable Place Branding." In *International Place Branding Yearbook 2012: Managing Smart Growth and Sustainability*, edited by Frank M. Go and Robert Govers, 147. New York: Palgrave Macmillan.

Schilling, Joseph, and Jonathan Logan. 2008. "Greening the Rust Belt: A Green Infrastructure Model for Right Sizing America's Shrinking Cities." *Journal of the American Planning Association* 74 (4):451–466.

Schilling, Joseph M., and Alan Mallach. 2012. *Cities in Transition: A Guide for Practicing Planners*. Vol. 568, *Planning Advisory Service*. Chicago: American Planning Association.

Silverman, Robert Mark, L.I. Yin, and Kelly L. Patterson. 2013. "Dawn of the Dead City: An Exploratory Analysis of Vacant Addresses in Buffalo, NY 2008–2010." *Journal of Urban Affairs* 35 (2):131–152.

Swyngedouw, Erik, Frank Moulaert, and Arantxa Rodriguez. 2002. "Neoliberal Urbanization in Europe: Large Scale Urban Development Projects and the New Urban Policy." *Antipode* 34 (3):542–577.

Teaford, Jon C. 2000. "Urban Renewal and Its Aftermath." *Housing Policy Debate* 11 (2):443–465.

Tiebout, Charles M. 1956. "A Pure Theory of Local Expenditures." *Journal of Political Economy* 64 (5):416–424.

Weaver, Russell, and Chris Holtkamp. 2015. "Geographical Approaches to Understanding Urban Decline: From Evolutionary Theory to Political Economy . . . and Back?" *Geography Compass* 9 (5):286–302.

Weber, Rachel. 2002. "Extracting Value from the City: Neoliberalism and Urban Redevelopment." *Antipode* 34 (3):519–540.

Wood, Gordon S. 1992. *The Radicalism of the American Revolution*. New York: Alfred A. Knopf.

7 Rightsizing and smart decline

In Chapter 6, we explored some of the underlying assumptions that exist in American urban policy. We begin here by noting that, at the most basic level, decision-makers in shrinking cities face a choice between taking action or not acting in response to persistent population loss (Hollander and Cahill 2011). Hospers (2014) expands this set of choices into a four-way **typology of policy responses to urban shrinkage**. The four types of policy responses are:

1 *Trivializing urban shrinkage*: the "do nothing" or "no action" option in which policymakers discount the seriousness of urban shrinkage and stand by the status quo package of policies.
2 *Countering urban shrinkage*: the *pro-growth* response (Ch. 6) in which actions are taken to foster urban growth through attracting new residents and external private investment.
3 *Accepting urban shrinkage*: the option in which decision-makers openly admit to the improbability of rapid re-growth and shift their attention to actions that stabilize population and improve conditions for remaining residents.
4 *Utilizing urban shrinkage*: presumably a by-product of the *accepting* option, shrinkage is positively framed as a unique and promising opportunity to experiment with new urban fabrics that can better serve smaller, if not still declining, populations.

Of these responses, *trivializing* and *countering* seem to be the most frequently encountered in the United States (Hackworth 2014, 2015). This observation is directly related to the pervasiveness of *pro-growth approaches* in American policy-making arenas (Ch. 6). Nevertheless, *accepting* and *utilizing* shrinkage are rapidly gaining momentum and followers in the shrinking cities' planning and research communities, as stakeholders from all sectors of society are coming to terms with persistent population shrinkage (Popper and Popper 2002). Indeed, the movement to acknowledge and embrace shrinkage has led to, and continues to produce, novel theoretical and practical frameworks for "decline-oriented" (as opposed to traditional growth-oriented) urban planning and policy (Schatz 2012).

In that vein, the remainder of this chapter outlines several inter-related concepts that are consistent with *accepting* and/or *utilizing* shrinkage. Drawing on current research, real-world examples, and emerging trends, we describe how decision-makers in shrinking cities are both envisioning and attempting to create the conditions for an urban environment in which "small is beautiful", and economic *growth* is subordinate to socio-ecological *development* (Popper and Popper 2002; Axel-Lute 2007; Ch. 6). After providing rather broad introductions to relevant mental models, our attention quickly turns to investigating viable policy instruments and strategies that can be adopted in their pursuit. We conclude the chapter by offering a reminder that paradigm shifts (i.e., transformations of mental models) require time, commitment, and – as we pick up with in Chapter 8 – often involve attendant changes to the system.

From a shrinking municipality's point of view, the reasons for accepting and making the most of smaller population levels are manifold, though troubling for decision-makers in many American cities. Namely, cities are conventionally seen as paragons of **economies of scale** (Batty 2013). This idea relates to the straightforward premises that the more people there are in a fixed set of municipal boundaries, and the denser the corresponding settlement patterns, the lower is the cost per person of delivering services to a given municipal population. Relatedly, a larger resident population implies more municipal property tax revenue with which to deliver public services. When population rapidly thins out, as it has in shrinking cities, municipalities are left with fewer resources with which to provide the same assortment and quality of services across the same spatial extent of their [now de-densified] urban landscapes. The result is that for many cities that have experienced substantial population loss, it becomes more and more cost prohibitive for them to offer the full portfolio of services they were once capable of providing the larger, denser population. Thus, in some cities, municipal leaders have begun to create frameworks for cutting back the facilities they provide to predominantly de-settled areas of the city.

Notably, proposals geared toward *accepting* shrinkage are sometimes bemoaned for two seemingly opposite reasons. On one hand, accepting shrinkage is akin to admitting "failure" in the neoliberal American model of urban development (Leo and Anderson 2006; Ch. 6). As a consequence, public officials and policymakers often oppose decline-oriented planning. On the other hand, there are vitally important social justice concerns – a topic that is discounted or even overlooked in the American model of urban development (Kantor 2010) – that come along with potentially cutting off services to sectors of cities. Even if relocation strategies are put forth to move residents from shrinking and/or declining areas into better quality housing units in denser parts of the city, residents may wish to remain in their existing homes for a variety of reasons (Hollander and Németh 2011). Hence, embracing shrinkage as a means for consolidating population and re-establishing economies of scale tends to be a subject of debate across the political spectrum.

While the notion of urban consolidation is not the matter at hand – it will be revisited later in the chapter – introducing it at this juncture is helpful for conceptualizing what is arguably the core challenge in developing alternative approaches: spatial and scalar *mismatches* between the built and social subsystems of shrinking cities. The term 'scale' is used here in the sense of relative size – what matters is the geographic extent (size) and spatial distribution of people *relative to* the geographic extent (size) and spatial distribution of the built environment.

When the size of these built and social subsystems are approximately balanced, cities exhibit the **economies of scale** for which they are known (Batty 2013). A **scale mismatch** occurs between urban social and built subsystems when the scale of the built subsystem and the scale of the social subsystem responsible for managing it are "aligned in such a way that one or more functions of the [overall urban] system are disrupted, inefficiencies occur, and/or important components of the system are lost" (Cumming, Cumming, and Redman 2006: 3). In the case of shrinking cities, extensively de-densified social subsystems, which are the products of decades-long population loss, have become decidedly mismatched to the comparatively large-scale built subsystems they are responsible for managing and maintaining (Glaeser and Gyourko 2006). As the foregoing definition suggests, this situation has resulted in, among other issues, inefficiencies in public service provision and the disruption of desirable urban systemic functions – especially those functions that relate to healthy real estate market demand and neighborhood upkeep (Schilling and Mallach 2012). Consequently, the following conceptual and policy frameworks that *accept* and/or *utilize* shrinkage devote considerable attention to developing strategies for bringing scale-mismatched social and built subsystems into a more balanced relationship.

Smart decline

The discipline of planning emerged, and largely continues to operate, as a means for controlling the rate, geographic distribution, and consequences of urban growth. In this context, scholars observe that the traditional tools of [classic, growth-oriented] planning are poorly suited to the circumstances of shrinking cities (Schilling and Mallach 2012). Responding to this incompatibility between tools and conditions, Popper and Popper (2002: 21–22) challenged practicing and academic planners to develop an alternative approach that "leaves behind assumptions of growth". They refer to this approach, which strives to improve quality of life for all remaining citizens rather than attempt to attract new ones, as **smart decline**.

The label *smart decline*, or, as it sometimes called, "smart shrinkage" (Rhodes and Russo 2013), is a play on the popular Smart Growth movement in urban planning (Axel-Lute 2007; see Daniels 2001 for an overview of Smart Growth). As Warner (2006: 169) describes it:

> [t]he basic idea of Smart Growth is that growth should occur within or immediately around already existing urban areas. Smart Growth can allow communities to preserve open space, natural areas, and farmlands; maintain historic investments in cities; develop attractive, compact metropolitan areas with a decreasing emphasis on the automobile; create mixed-use neighborhoods so that people can walk to work, shopping, and entertainment; and maintain the unique character of neighborhoods and towns. Smart Growth's antithesis is sprawl.

Just as Smart Growth calls for the creation of compact urban forms from dense, mixed-use urban fabrics, smart decline is about re-establishing economies of scale in shrinking cities through denser and more vibrant settlement patterns that enhance the quality of life for current citizens. Unlike Smart *Growth*, however, smart decline does not anticipate and lay the groundwork for absorbing future population increases. Instead, smart decline means "planning for less – fewer people, fewer buildings, fewer land uses" (Popper and Popper 2002: 23). Thus, to achieve the denser, higher-quality settlement patterns that the smart decline mental models recommend, strategies and actions for balancing the scales of the social and built subsystems are necessary.

Numerous such strategies and actions have been proposed or attempted, and many of them will be discussed momentarily. For now, it is helpful to start by quickly summarizing a purported "foundational theory for planning shrinking cities" from the smart decline literature. Explicitly, Hollander and Németh (2011) put forward a theory (more accurately, a set of criteria) that seeks to define the parameters of a smart decline planning process. Grounded in scholarship on procedural social justice, the essence of the theory can be conveyed through five normative propositions (Hollander and Németh 2011: 358–361):

1 Smart decline planning processes must include and explicitly recognize multiple voices.
2 Smart decline planning processes should be political and deliberative in nature.
3 Smart decline planners should be cognizant of differential communication techniques and should provide information that enables citizens to recognize and challenge power imbalances and structures of domination.
4 Smart decline planning processes must be transparent and value different types and sources of information.
5 Smart decline planning processes should be regional in scope but local in control and implementation.

Separately, each of the above propositions concerns the manner by which a particular component of a planning process – e.g., the actors, interactions, roles, decision-making rules, or methods and monitoring procedures – is operationalized. Viewed as a collective set of premises, however, they imply that smart decline is an outgrowth of **communicative planning theory** (CPT). In brief, CPT is a reaction to **rational planning theory** (RPT). Within RPT, trained "experts" draw on empirical data, in conjunction with "techno-scientific analysis and deductive logic", to make planning choices on behalf of a community or city with little concern for local or experiential knowledge (McGuirk 2001: 196). CPT, on the other hand, calls for active citizen participation, such that the futures of communities are "deliberated and debated in public, and decided in democratic and inclusive processes" (Sager 2011: 181). CPT, therefore, seeks out and values knowledge of all types, especially local knowledge, and does so in such a way that planners and citizens are active partners as opposed to agents and principals. Whereas RPT is the prevailing approach found in most top-down neoliberal planning regimes (Sager 2011), CPT is viewed as the "best practice" among most bottom-up community planning practitioners and organizations. In this regard, smart decline theory implies that re-aligning the scales of the built and social subsystems of shrinking cities demands a substantial amount of bottom-up action.

Rightsizing

Re-conceptualizing efforts to balance the scales of the built and physical subsystems of shrinking cities as **rightsizing** is often credited to Schilling and Logan (2008) and their proposal to convert underutilized parcels in depopulated sectors of cities into "green" infrastructure and land uses (discussion to follow). While the literature on rightsizing *per se* has outwardly focused less on theory and process compared to the literature on smart decline and has been more geared toward actions and outcomes (Hollander and Németh 2011; Mallach 2011), current conventional wisdom holds that the two terms are synonymous (Axel-Lute 2007). However, many scholars now show a preference for the comparatively uncharged label of "rightsizing" over the somewhat more equivocal, and potentially pejorative, "smart decline". As Mallach (2011: 372 [note 8]) points out in a critique of Hollander and Németh's (2011) "foundational theory" (see above), *smart decline* sends a message that cities adopting this mental model wish to continue declining, when, in fact, most aim to stabilize. Accordingly, the remainder of this chapter will employ the term *rightsizing* when there is a choice between the two labels (as in, for instance: Beauregard 2012; Ryan 2012; Hummel 2015).

Selected rightsizing strategies and policy instruments

This section provides a general introduction to several of the most popular and influential rightsizing policy instruments from the shrinking cities planning literature. Following Oswalt (2006), the policies are grouped into four broad classes of strategies: (1) disassembling[1]; (2) re-evaluating; (3) re-organizing; and (4) imagining. To the extent that a given policy instrument takes on elements of more than one of these [non-mutually exclusive] categories, it is grouped with the class to which it is deemed to be most closely related. Once the essential (general) properties of the selected policy instruments from each of the aforementioned categories are outlined, real-world examples from two shrinking Ohio cities – Youngstown and Cleveland – are drawn upon to explore how rightsizing policies are implemented and how they play out in practice.

Disassembling

According to Oswalt (2006: 18), *disassembling* strategies are adopted to "roll-back" or "de-urbanize" city space, specifically by manipulating the physical urban fabric (Hollander and Cahill 2011). From the perspective of rightsizing, disassembling policy instruments are tools for subtracting excess infrastructure and buildings from the physical landscape for the purpose of bringing the social and built subsystems of a shrinking city into a more balanced and more functional relationship. Three prominent policy instruments that fit into this class of rightsizing strategy are *demolition*, *deconstruction*, and *consolidation*.

Demolition

Two of the most visible correlates of urban shrinkage are vacant and abandoned property and real property disinvestment. These phenomena contribute not only to perceptions of urban blight and negative environmental images (Bales 1985) but also to local government fiscal crises (Mallach 2012). As already discussed, the continued provision of municipal services and infrastructure maintenance to depopulated parts of the city is cost prohibitive. In most cases of pronounced population loss, economies of scale for these activities break down. Moreover, vacant and blighted properties typically require costly government investments that extend beyond standard infrastructure and service provisions, including increased attention to building (re)inspections and code enforcement, graffiti removal, and related reactive measures (Rhodes and Russo 2013). Thus, these so-called "nuisance" properties are popular topics of conversation in the rightsizing discourse (Mallach 2010; Frazier, Bagchi-Sen, and Knight 2013).

Perhaps the most common and widely utilized tool for confronting the vacant property issue in shrinking cities is structural *demolition* – that is, tearing down buildings from the physical urban fabric that are judged by local stakeholders to be deleterious. Chapter 6 argued that large-scale demolition programs have repeatedly been used in U.S. shrinking cities for the primary purpose of clearing land for economic growth. Despite this popular critique (Ryan 2012), demolition is a necessary component of serious rightsizing efforts. Mallach (2012) makes this case on two grounds. First, the macro justification is that the supply of housing units in shrinking cities greatly exceeds demand. Many areas of shrinking cities are characterized by "weak markets" or "dead zones" in which real estate investment is negligible or nonexistent (Schilling and Mallach 2012). This observation has two crucial implications: (1) the majority of vacant and abandoned properties in shrinking neighborhoods are likely to remain uninhabited for the foreseeable future; and (2) combined with the reality that vacant and abandoned properties are often subject to disinvestment and poor structural conditions, rehabilitating them rarely produces economic returns in the real estate market (Mallach 2012, 9). Accordingly, demolition is seen as the most viable option. Second, the micro justification concerns the extent to which vacant and abandoned properties undermine quality of life and social relations in the communities where they are located (Frazier, Bagchi-Sen, and Knight 2013). Blighted structures devalue nearby properties, which can trigger feedback effects whereby other property owners in the neighborhood decrease their own investments into maintaining the appearance of the shared "urban commons" (Weaver 2015). To prevent such outcomes, then, ridding neighborhoods of nuisance properties is a reasonable option (Mallach 2012: 13).

Table 7.1 Mallach's (2012) ten action steps for strategic demolition in shrinking cities

1.	Adopt transparent and efficient procedures to evaluate which buildings are to be demolished.
2.	Establish priority criteria for demolition, and strategically target neighborhoods where demolitions are likely to have the largest gains (as opposed to ad hoc demolition, which is commonplace in many shrinking cities [Ryan 2012a]).
3.	Democratize the demolition decision-making process to include the widest range of interests and viewpoints possible.
4.	Adopt efficient procedures to gain legal authority to demolish privately owned buildings and take title to vacant buildings and lots.
5.	Incorporate specific steps to prevent resulting vacant lots from becoming sources of new urban blight (e.g., overgrowth, illegal dumping, etc.).
6.	Develop integrated neighborhood stabilization programs where demolition is linked to ongoing or planned rehabilitation and reuse efforts.
7.	Work with state governments to review state statutes and regulations affecting demolition, and modify or abolish those rules that impose unreasonable costs on demolition efforts.
8.	Use state legal tools to recover costs of demolition and advocate for strong state support of demolition.
9.	Work with state officials to support federal neighborhood stabilization programs.
10.	Commit state resources to local demolition efforts.

In this light, it is undeniable that rightsizing shrinking cities calls for the demolition of excess abandoned and functionally obsolete buildings (Hummel 2015). The challenge, however, is to use demolition strategically, in ways that do not hinge on the (unlikely) prospect of future economic and population growth (contra Ch. 6). Instead, demolition activity must be executed in support of remaining residents' quality of life (Frazier and Bagchi Sen 2015). To that end, Mallach (2012) offers ten action steps for local governments (Table 7.1). Similar to the implications of Hollander and Németh's (2011) "foundational theory" that was summarized above – though distinct in their prioritization of outcomes over process (see Mallach 2011 on this distinction) – these action steps suggest that rightsizing is more closely aligned with communicative planning theory than rational planning theory.

The problems of widespread property vacancy and abandonment, disinvestment, and urban blight are examples of *cumulative causation*. Recall from Chapter 1 that cumulative causation refers to the idea that once a negative qualitative change has started in a given area, it is reinforced by a positive feedback. This feedback effect works to intensify the rate at which the negative change occurs in each time period after its introduction (Hospers 2014). While the incidence of property neglect and abandonment during a given time interval is influenced by any number of exogenous factors – e.g., job loss, stock market crash, serious illness or death, fire or natural disaster, etc. – the presence of abandoned and blighted property in a neighborhood is self-reinforcing. Blight and abandonment devalue nearby property, which discourages erstwhile responsible owners from reinvesting

Table 7.2 Vacancy and abandonment in selected cities, 2000–2013

City	*Vacant units, 2000 (% of total units)*	*"Other" vacant, 2000 (% of vacant units)*	*Vacant units, current (% of total units)*	*"Other" vacant, current (% of vacant units)*	*Change in vacancy rate*	*Change in "other" vacant (% of vacant units)*
Baltimore, MD	14.1%	49.4%	18.5%	58.5%	4.4%	9.1%
Buffalo, NY	15.7%	43.7%	16.9%	68.9%	1.2%	25.2%
Cleveland, OH	11.7%	32.9%	13.8%	64.1%	2.1%	31.2%
Detroit, MI	10.3%	43.7%	29.3%	65.5%	19.0%	21.8%
Flint, MI	12.1%	30.8%	23.6%	72.0%	11.5%	41.2%
Gary, IN	12.3%	47.9%	25.6%	89.4%	13.3%	41.5%
Johnstown, PA	13.0%	31.9%	19.4%	72.3%	6.4%	40.4%
New Orleans, LA	12.5%	39.9%	21.9%	51.4%	9.4%	11.5%
St. Louis, MO	16.6%	44.0%	19.9%	61.3%	3.3%	17.3%
Youngstown, OH	13.4%	20.8%	20.2%	57.3%	6.8%	36.5%

Sources: U.S. Census 2000 SF 1, Tables H3 and H5; U.S. Census American Community Survey 5-Year Estimates (2009–2013), Tables B25002 and B25004; NB: Johnstown, PA is the only city in the table that is not included in our list of shrinking places from Table 2.3.

in their properties. As maintenance standards begin to fall (disinvestment), more owners eventually abandon the affected neighborhood (National Vacant Properties Campaign 2005). Evidence suggests that property blight and abandonment do tend to exhibit this sort of behavior.

In Buffalo, New York, for example, an empirically detected cluster of substandard property conditions in the central part of the city was found to grow in size over a ten-year period, as blight seemed to spill over from the original cluster into spatially contiguous areas over time (Weaver and Bagchi-Sen 2013). At a more aggregate level, Table 7.2 presents U.S. Census Bureau housing vacancy data for ten shrinking cities at two points in time (2000 and the period from 2009–2013). Importantly, census vacancy data are classified into the following types: *for sale*, *for rent*, *sold*, *rented*, *seasonal or recreational*, *migrant worker housing*, and *other*. The latter of these categories, *other*, describes housing units that are vacant for reasons other than being on the market or having a dedicated seasonal use. As a consequence, this category of vacancy includes "those units that are abandoned and blighted" (Schilling and Logan 2008). Table 7.2 shows the overall vacancy rate for each shrinking city along with the percentage of vacant units classified as *other*. Note that for all cities listed in the table, both of these quantities increased between 2000 and the present. While, again, there are many exogenous factors at play in the vacancy and abandonment process, the data in Table 7.2 almost certainly reflect the aggregate results of neighborhood-level positive feedback operating on property disinvestment.

Deconstruction

In contrast to demolition, during which structures are torn down and their building materials are discarded, *deconstruction* is defined as the "careful or systematic dismantlement of buildings in such a way that the individual building components are separated and preserved for potential reuse" (Mallach 2012: 21). Deconstruction is an appealing alternative to demolition when targeted buildings are historically significant or otherwise possess distinctive material features. Deconstruction is also typically favored over standard demolition from an environmental perspective. As nonprofit deconstruction organizations such as Buffalo ReUse (in Buffalo, New York) point out, sending salvageable material to a landfill – the standard procedure of ordinary demolitions – is both wasteful and oftentimes destroys "architectural gems" (Buffalo ReUse n.d.). On the other hand, deconstruction significantly increases the cost of tearing down a building, as it is invariably more labor-intensive than traditional demolition. Furthermore, there is a limited market for deconstruction products. For these reasons, Mallach (2012) suggests that deconstruction should remain an option in shrinking cities but be approached on a case-by-case basis. When the costs of deconstruction less the value of salvaged materials still exceeds the cost of traditional demolition, the latter is the more efficient and expedient option.

Consolidation

The farthest-reaching and most controversial of the *disassembling* rightsizing policies discussed in this subsection is *consolidation*. The basic notion of consolidation is for the city, as a consensual venture of citizens and decision-makers, to effectively dispossess the infrastructure and buildings located in predominantly desettled neighborhoods. Stated more simply, the idea is to stop providing municipal services to and maintaining/investing in infrastructure in predominantly depopulated parts of the city (Hummel 2015). Over time, the buildings and infrastructure in these neighborhoods can be liquidated – demolished, deconstructed, or functionally abandoned – to make room for alternative land uses, such as re-naturalization. In the immediate term, the city's relatively viable (i.e., denser) neighborhoods would act as receivers for the residents of soon-to-be decommissioned sectors of the city. Public resources would be made available to facilitate this resident relocation / neighborhood consolidation process, which would simultaneously increase the density of healthy neighborhoods, decrease municipal service provision costs, and, putatively, enhance the quality of life for relocated residents who leave behind conditions of disinvestment and blight for more stable residential environments (Hackworth 2015).

On this backdrop, *consolidation*, combined with strategic *demolition* and *deconstruction*, appears to offer one of the most expeditious and expedient paths available for addressing the *scale mismatch* between the built and social subsystems of shrinking cities. Note well, though, that the ideal description of a consolidation process from the preceding paragraph does not generally represent real-world conditions. Almost always, cities are constrained in their abilities to undertake comprehensive consolidation efforts by lack of financing and insufficient higher-level (e.g., state and federal) support (Axel-Lute 2007; Rhodes and Russo 2013). Moreover, consolidation, which requires decision-makers to identify viable and non-viable neighborhoods, can function in practice more like rational than communicative planning – where the latter is viewed as the appropriate framework for rightsizing (Hollander and Németh 2011). The reason for this discrepancy comes from the fact that non-viable neighborhoods are often classified as such on the basis of various economic, physical, and demographic indicators, and this classification requires (expert) analysis and interpretation of empirical data.

The risk of such an approach is to mute the voices of residents in the neighborhoods targeted for consolidation (Axel-Lute 2007). If such residents do not wish to relocate, and in turn become holdouts in consolidation efforts, then cities face the dilemma of using their **eminent domain** powers to take title to holdouts' properties (in return for just compensation and, it follows, relocation to another part of the city). Even though eminent domain takings are legal when they serve a "public purpose" – and the long-term benefits of consolidation would almost surely qualify as a valid public purpose (LaCroix 2011) – the unwanted taking of property, regardless of the compensation, does not align with most understandings of social justice. Consequently, if a city elects to pursue consolidation, then considering it in partnership with an active citizenry, within a communicative planning process, may help minimize the potentially harmful outcomes that are possible under this policy proposal (Hollander and Németh 2011).

Re-evaluating strategies

Whereas the preceding strategies are concerned with the *dis*use of urban elements that are mismatched to their current contexts in shrinking cities, *re-evaluating* strategies consider the *re*use of those elements (Oswalt 2006). It is generally recommended that *re-evaluating* strategies accompany *disassembling* strategies, and that the two act in concert to bring about quality-of-life-enhancing outcomes (e.g., Table 7.1). In other words, infrastructure and buildings should only be demolished or abandoned (with the exception of cases that are clear and immediate threats to public safety) if a reuse has been planned and agreed upon for the target sites (Mallach 2012). Within the shrinking cities literature, the types of reuse/re-evaluating strategies that are most frequently proposed go by many names – "urban greening" (Hummel 2015), "green infrastructure" (Schilling and Logan 2008), and "green land uses" (LaCroix 2011), among others – but all of these names point to a common approach: *downzone* intensely used spaces to accommodate less environmentally impactful activities, and, in the process, *de-densify* the urban landscape.

Downzoning

Downzoning occurs when a tract of land that was previously designating for relatively intense activity is legally rezoned to a comparably restrictive, lower impact use (Kuhn 2011). A current mainstream idea among shrinking cities scholars is to downzone mostly abandoned residential, commercial, and industrial districts to one or more "green" uses, including large-scale agriculture, community gardens, dedicated open space, alternative energy production, and urban forests or other forms of re-naturalization (Schilling and Logan 2008; LaCroix 2011; Lawson and Miller 2012; Frazier and Bagchi-Sen 2015).

LaCroix (2011) observes that downzoning for these types of "green" uses is within the power of local governments (Ch. 7), as long as the reasons for the changes are non-arbitrary, non-capricious, and enacted in the public interest. However, there are a number of practical issues to consider. First, downzoning is commensurate with, and in many ways a logical implication of, *consolidation*. If sectors of a city are targeted for consolidation, then any resultant disused spaces cannot simply be forgotten; for leaving a crumbling physical fabric in place poses threats to public health and safety (LaCroix 2011). An appealing alternative is therefore to demolish or deconstruct the infrastructure and buildings in consolidated spaces, and subsequently downzone those areas to one or more of the aforementioned green uses (Schilling and Logan 2008). To the extent that consolidation plans are met with uniform acceptance and sufficient funding, this marriage of disassembly and downzoning is a happy state of affairs.

More realistically, however, there are likely to be holdout stakeholders who object to the proposed changes. In these circumstances, should the government still proceed with downzoning, holdout residents may legally challenge this action on the grounds of a **regulatory taking**. Related to the notion of *eminent domain* (see above), a regulatory taking occurs when a government regulation is so onerous that it reduces the economic exchange value of a given property. That is, if downzoning proscribes future, say, commercial activity in a newly established green zone, then any existing commercial structure already in that space is plausibly worth less (if not *worthless*) under the new zoning rules. Thus, as was the case with *consolidation*, *downzoning* efforts that occur as part of a communicative planning process may minimize potentially harmful outcomes. Such a process can help decision-makers discover stakeholder sentiments and preferences before regulatory change processes are undertaken.

Second, as the preceding paragraph suggests, downzoning large tracts of land is a delicate and challenging task, as it often involves a large number of heterogeneous stakeholders. Accordingly, governments might instead elect to downzone on a parcel-by-parcel basis. In these scenarios, local authorities are vulnerable to challenges of **spot zoning**. Spot zoning occurs when the local government enforces its land use regulations unequally (Fischel 1987). For example, suppose that an existing residential parcel is re-zoned after a structural demolition to allow light agriculture on the premises. If, after this change, a homeowner on the affected block is then punished by the government for engaging in light agricultural activity on his or her own residential property (e.g., in the backyard), then such circumstances might qualify as unequal treatment in the eyes of the law. Importantly, though, not all spot zoning is illegal, and parcel-by-parcel downzoning decisions can be upheld as long as they are justified by some guiding policy or planning framework (LaCroix 2011).

Third, several of the "green" land uses proposed for shrinking cities have the potential to unduly burden neighboring citizens or landowners. For instance, where agriculture is authorized in close proximity to existing urban residential uses, the odors, noises, and other products of agricultural activity (e.g., pesticide drift) are almost certain to affect the quality of life of neighboring residents. Similarly, downzoning for alternative energy production, such as wind turbines, has the potential to disrupt extant viewsheds and increase noise levels in a targeted community (LaCroix 2011).

Finally, and perhaps most seriously, many of the most abandoned parts of shrinking cities are often located on or near former industrial sites that, in all likelihood, contain hazardous substances, pollutants, or other contaminants (Schilling and Mallach 2012). These **brownfields** – i.e., parcels of land that require clean up and remediation prior to their reuse (U.S. Environmental Protection Agency n.d.) – are therefore simultaneously some of the top candidates for downzoning given their level of abandonment, and some of the costliest places to implement downzoning given the financial capital needed to finance necessary remediation.

These concerns add up to suggest that downzoning is much more complex than merely redrawing boundaries on a map (see Register 2006). Rather, like the other tools of rightsizing discussed to this point, downzoning calls for a holistic, communicative planning process that merges technical expertise and data analysis with knowledge of community-level priorities and visions. It further calls on municipalities to engage directly with longstanding issues of environmental contamination from past industrial activity. These issues will undoubtedly require outside assistance from higher levels of government, given the costs involved. Assuming that such challenges are surmountable, however, downzoning depopulated neighborhoods to "green" uses represents an exciting and potentially transformative option for rightsizing shrinking cities (Schilling and Logan 2008).

De-densification

De-densification is effectively the inverse of *consolidation*. Rather than emptying out some parts of the city and clustering population in others, thereby weaving together a patchwork of vacant and high-density neighborhoods, *de-densification* fosters population dispersion (Hollander et al. 2009). Together with demolition and/or deconstruction, de-densification policies encourage urban property owners to take title to vacant parcels adjacent to their lots. The rationale is that by extending ownership of previously vacant or blighted properties to the remaining members of a depopulated area, at low costs or through other incentives, de-densification policies empower local residents to meaningfully reshape their neighborhoods through bottom-up action (Lawson and Miller 2012). A corollary of this reasoning is that average lot sizes in the city will increase and, by extension, residential density will trend downward across the municipality.

Overall, the focus of de-densification policies on bottom-up neighborhood action and empowering local residents ostensibly adheres to the communicative planning framework advocated for by rightsizing scholars (Hollander and Németh 2011). Further, the small-scale, incremental approach of de-densification suggests that these so-called "side lot" programs, which are a specific form of downzoning, can quickly adapt to and enhance livability in affected neighborhoods (Marshall 2009; Weaver and Holtkamp 2015). However, in practice, it has been found that "restrictive guidelines and inequitable or illogical pricing structures" have undermined the success of de-densification programs (Ganning and Tighe 2014: 1). The implication is that such policies might benefit from a participatory public audit process to identify their weaknesses, recommend changes, and ultimately improve their efficiency and functionality. Furthermore, the success of de-densification programs hinges on the relative number and spatial distribution of vacant and abandoned properties. Remaining residents are only able to absorb and maintain a certain amount of additional property. When the ratio of abandoned to remaining properties reaches a certain threshold, de-densification strategies are not likely to succeed.

Reorganizing

The third broad class of rightsizing strategies, *reorganizing*, deals with questions of social organization and urban governance (Oswalt 2006). As Hollander and Cahill (2011: 255) note, reorganizing is more about manipulating the management structure of a shrinking city as opposed to its physical elements. In addition to the seemingly necessary shift from rational to communicative planning in shrinking cities (Hollander and Németh 2011), and a transformation in mental models from *countering* to *accepting* and/or *utilizing* shrinkage (refer to the *typology of policy responses to urban shrinkage*), several more specific reorganizing tools and ideas have been proposed in the literature. This section focuses on *place-based palliative planning interventions*, *building social capital*, *alternative ownership models*, and *polycentric governance*

Place-based palliative planning interventions

The rise of neoliberal ideology in urban America was accompanied by a disenchantment with public sector intervention in city-building and (sub)urban development (Knox 2011; Ch. 6). Far from the place-makers that they were during the urban renewal era of the 1950s and 1960s, neoliberal-era local governments have become place-*marketers* that aim to (1) create conditions that are attractive to outside investors and (2) allow patterns of private market investments to dictate the trajectory of a city's development (Madanipour 2006). *Place-based palliative planning interventions* represent a proposed *reorganization* of this market-first style of local government, in that they call for public authorities to once again operate as place-makers and active interveners in the management of urban change (Ryan 2012).

Ryan (2012) offers that *demolition* programs, especially those that operate at a large scale, are one form of place-based, interventionist policy. However, in the absence of *reevaluating* strategies, this type of intervention can be perceived as negative or reactive. In contrast, carefully crafted and strategically targeted *positive* or *proactive* interventions, such as constructing new public facilities or engaging in local beautification, are plausibly where *palliative interventions* can have the most impact (Ryan 2012: 206). For example, Savitch (2011) relates evidence from Marseille, France, in which high-intensity concentrated development in a key geographic sector of the city, along with significant citywide investments in public transit, came together to produce transformational changes – including repopulation and job creation – in what had been a shrinking city.

At a much smaller scale, Weaver and Knight (2015) analyzed a palliative intervention in Buffalo, New York, referred to as the "clean sweep" program. This policy initiative is a collaboration of more than a dozen municipal agencies that "descend on neighborhoods" to (1) improve public spaces through small-scale [visible] projects such as litter removal and (2) actively engage residents in conversations about their communities through door-to-door interactions (Olavsrud 2013). During these conversations, public officials disseminate information about essential public services, and they attempt to bring residents together to form *neighborhood associations* – groups that are formally recognized by the city, are consulted during official planning processes, and meet regularly to discuss local issues. Using data from the Buffalo community in which this "clean sweep" program was most active during its first full year of implementation, Weaver and Knight (2015) found that properties located on blocks where clean sweeps were performed were significantly less likely to commit property code violations in the following year, relative to a control group of properties in the same community.

Building social capital

The Buffalo "clean sweep" initiative, apart from being a *place-based palliative intervention*, is also a convenient example of a tool for building social capital (recall that part of the program attempted to bring residents together in neighborhood associations). As Chapter 4 argued, group-level social capital can be operationally defined as the "capacity of a [neighborhood] for collective action" (Turchin 2003: 43). Research has shown that citizens in neighborhoods with high social capital are routinely able to work together for the good of their group (e.g., Ahn and Ostrom 2008) and are often capable of collectively resisting negative qualitative changes (Temkin and Rohe 1998). Put differently, social capital appears to be a mechanism for bottom-up neighborhood self-governance. The organic and emergent qualities of outcomes that are produced by this type of collective action are therefore attractive both from a communicative planning perspective and from a public finance perspective. For these reasons, scholars have made convincing arguments that building social capital is a potential solution to the vacant land problem in shrinking cities (Nassauer and Raskin 2014). Yet, notwithstanding the theoretical appeal of social capital, the success of community-based collective action largely depends on the degree to which neighborhoods are able to make at least some of their own rules and decisions, and have the authority to conduct their own affairs (Wilson, Ostrom, and Cox 2013; also see Ostrom 1990, 2005, 2009). In this sense, building social capital is not sufficient – the environment must also be right for social capital to function properly. The last two *reorganizing* tools discussed here engage with this issue more directly.

Alternative ownership models

The prevailing view of property rights in the United States, referred to as the **ownership model** (Blomley 2004), holds that property is almost exclusively *private*. That is, for each parcel of real property, there is a clear, identifiable owner. Above all, this model recognizes that owners have the right of **alienability**, which means that owners are free to sell their property to willing buyers. Many scholars in the United States assert that without alienability, users of property effectively have *no* property rights at all (see the discussion by Ostrom 2009). Thus, in the ownership model, there are private landholders who can sell their land as they wish; and there are public landholders who possess the same right – though the former is viewed as normative ownership where the latter is seen as "inferior" (Blomley 2004: 7).

Lawson and Miller (2012) observe that this narrow understanding and interpretation of ownership undermines and disincentivizes collective, bottom-up efforts to improve conditions in distressed urban neighborhoods. Because taking title to abandoned property tends to be costly and time consuming in many U.S. cities (Ganning and Tighe 2014), community members occasionally engage in maintenance of abandoned properties or vacant lots without legal authority to do so (Dewar and Thomas 2012). Under the ownership model, these *community-based initiatives* (see Ch. 6) are inherently risky investments: those that are well-executed and succeed in creating new value in their neighborhoods are vulnerable to situations in which legal deed holders (including the local government) exercise their rights of alienability over the now more valuable properties (Lawson and Miller 2012: 18). Thus, there is a nontrivial possibility that residents who expend meaningful time and resources on creating, for instance, a community garden, will lose their full investment.

To protect communities and their social capital investments from such outcomes, scholars advocate for *alternative ownership models* in shrinking cities that recognize collective claims to property based on use and management, and that these considerations ought to trump legal ownership (i.e., alienability rights) in cases of abandonment (Blomley 2004; Lawson and Miller 2012). Ostrom (1990, 2009) observes that these types of "common property regimes" have succeeded in a wide assortment of contexts and flavors for long periods of time, and that they have done so in the absence of alienability rights.

Polycentric governance

Polycentric governance relates to the idea that human settlements contain many spheres of activity, and each sphere operates at a given level of organization. The task of implementing *polycentric governance* is thus to determine the operational levels for each activity sphere. From there, a general set of rules can be applied to all of these spheres by a central authority. Within each sphere, however, the global rules undergo localization and evolutionary processes that facilitate their adaptation to the heterogeneous contexts in which they are implemented (Ostrom 1972; Wilson 2014). Should these local rules result in intra- or inter-sphere conflict, the higher-level authority responsible for establishing the general (global) set of rules can intervene to resolve the conflict (Ostrom 2005).

One key insight from the polycentric governance literature is that functional institutions for managing collective action problems can, under the above arrangements, evolve from within a given sphere of activity (Ostrom 2009; Wilson, Ostrom, and Cox 2013). The lesson for shrinking cities is that empowering neighborhoods ("spheres of activity") with the authority to make and enforce some of their own rules can create conditions in which localized solutions to the vacant and abandoned property problem evolve from the bottom up. The benefits of bottom-up solutions are numerous: they tend to be better adapted than centralized rules to local circumstances; they are quicker to change than centralized institutions; and among other things, they allow for more experimentation and innovation than is possible under a single set of citywide regulations (Ostrom 2005).

Still, polycentric governance is not a panacea. The success of such systems hinges on the degree to which neighborhood boundaries can be drawn and community members are willing to participate in collective action (Ostrom 1990, 2005). Thus, establishing a polycentric governance structure would conceivably require a communicative planning process in which citizens help public officials delineate neighborhood boundaries. From there, policies such as the "clean sweep" intervention in Buffalo (see above) may be needed to *build social capital* in identified neighborhoods. While these are complex and delicate issues, inchoate evidence suggests that urban neighborhood governance arrangements are effective at preventing the rapid spread of blight (Weaver 2013) and other aspects of qualitative urban decline (Wagenaar 2007). As a consequence, neighborhood empowerment and polycentric governance can be important components of a holistic rightsizing plan in shrinking cities.

Imagining

The final category of rightsizing strategies, *imagining*, is the "most self-explanatory and hardest to operationalize" (Hollander and Cahill 2011: 255). In short, imagining involves re-conceptualizing the city – not necessarily as a place with an identity divorced from its heritage but one whose identity is detached from recent perceptions of shrinkage and decline. Along these lines, *imagining* is perhaps most apparent in future-oriented planning documents or vision statements. Two examples are the *Youngstown 2010 Citywide Plan* – which imagines Youngstown, Ohio, as a "better, smaller" city – and Cleveland's *Re-Imagining a More Sustainable Cleveland* – which lays out a series of specific goals for Cleveland, Ohio, to become a "greener", more environmentally conscious city. The core components of the Youngstown plan and Cleveland's vision statement are summarized below in Box 7.1 and Box 7.2, respectively.

Land banks and land trusts

In one way or another, all of the rightsizing policies discussed above are tools for reducing the *scale mismatch* between the social and built subsystems of shrinking cities. These policies were classified into four broad strategies: (1) disassembling; (2) reevaluating; (3) reorganizing; and (4) imagining. Crucially, two popular, but heretofore unmentioned, tools for rightsizing shrinking cities intersect all four of these strategy classes. Namely, *land banks* and *land trusts* are legal entities that take title to foreclosed properties (*reorganizing*), sometimes demolish structures on their land holdings (*disassembling*), regularly bundle properties into packages that are dispossessed to stakeholders who have reuse plans for them (*re-evaluating*), maintain lands as open space or re-naturalized habitat areas (*re-evaluating* – specific to land trusts), and, in performing these functions, contribute to the *re-imagining* of distressed neighborhoods and shrinking cities (LaCroix 2011).

Both **land banks** – the entities that acquire and assemble real property for eventual transfer and reuse – and **land trusts** – the entities that acquire real property to maintain it as conserved (e.g., "green") space – are therefore powerful institutions that shrinking cities can work with to manage their vacant and abandoned property problems. However, setting up these organizations, and funding them, generally requires authorization and support from state governments. Moreover, they are typically implemented at a county or other supra-local level, meaning that they necessitate *inter-governmental cooperation* (Hackworth 2014). For these reasons, land banks and land trusts are considered more extensively in the next chapter (Ch. 8), on metropolitan governance.

Box 7.1 Youngstown – a new vision

On September 19, 1977, a day locals still refer to as "Black Monday", Youngstown Sheet and Tube abruptly closed its Campbell Works plant, laying off 5,000 workers (Linkon and Russo 2002). Over the next five years, Youngstown Sheet and Tube closed another plant while U.S. Steel closed its Ohio and McDonald Works, and Republic Steel closed its Youngstown Works. In total, 50,000 jobs were lost regionally in steel and related industries, taking more than $1.3B in middle-class wages out of the economy and driving unemployment rates to more than 20 percent in the early 1980s. For Youngstown, the closure of so many steel plants compounded the impacts of suburbanization, which had been steadily drawing residents out of the city center. From 1950 to 1970, Mahoning County grew from 257,629 to 303,424 residents, an increase of 45,795, while the city of Youngstown lost

28,542. Population decline has remained persistent in Youngstown, dropping from 168,330 residents in 1950 to just 65,062 in 2014.

Physically, the city has suffered as well. In 1980, as steel plants were closing, Youngstown claimed 45,105 housing units with a vacancy rate of 6.9 percent. By 2010, the number of housing units had decreased to 33,123 (mainly through demolition), yet the vacancy rate had jumped to 19.0 percent. The implication is clear – the use of demolition to stabilize the vacancy rate during continued population decline has not been adequate. In 2009, a vacant-property survey conducted by the Mahoning Valley Organizing Collaborative estimated there were 4,500 vacant properties and 22,000 vacant lots, making 43.7 percent of Youngstown's parcels either vacant and/or abandoned (Tumber 2012). It is estimated that more than 130 properties are vacated each year, and by 2020, it is estimated there will be 7,500 vacant structures (Mahoning Valley Organizing Collaborative n.d.). Clearly, the problems in Youngstown mirror those of its shrinking city brethren. However, Youngstown made a key change in policy in 2005 when it adopted a new comprehensive plan that broke away from the pro-growth strategies commonly implemented in shrinking cities.

Youngstown 2010

Youngstown was the first city in the United States to formally adopt rightsizing principles in its master plan (Hackworth 2015). The plan, *Youngstown 2010 Citywide Plan* (hereinafter *Youngstown 2010*) began in 2002 as a collaborative endeavor led by the City of Youngtown and Youngstown State University, assisted by more than 200 volunteers, neighborhoods groups, and businesses, and with comments and feedback from more than 5,000 community members.

The *Youngstown 2010* plan created a "new vision for the new reality that accepts we are a smaller city" (City of Youngstown 2005: 7). Unlike other comprehensive plans common in declining cities, the acceptance of a smaller population and the intent not to formulate a plan based on models of economic and population growth was groundbreaking in public policy.

Much like Cleveland (see Box 7.2), Youngstown's leaders recognized the need to "re-imagine" the city and plan accordingly. To do so, the plan's strategies were guided by the following vision elements:

- The acceptance that Youngstown is a significantly smaller city and will remain so, at least in the near future. With that, the acceptance that the city is overbuilt with an overabundance of public infrastructure and buildings.
- Steel industries and manufacturing are not returning. Residents, leaders, and businesses recognize that a diverse economy must build on current strengths, including health care, higher education, and the arts.

- The city's image and quality of life are important and must be improved. The vision and image of the city is one of decay, which is a disincentive for residents and businesses and is also demoralizing to current residents. The city must focus on fixing the "broken windows" as well as address public safety, neighborhoods, downtown, and education.
- The people of Youngstown want to be involved and are ready for action. This vision, engagement, and community enthusiasm must be sustained, which requires an implementable, practical plan.

Fundamental to rightsizing policy in any shrinking city is how vacant land is to be managed, repurposed, and utilized. *Youngstown 2010* put forth land-use strategies based on four themes:

1 Transforming the city from "gray to green" by creating and expanding a network linking the city's existing green spaces and connecting the city's green network to the region.
2 Creating competitive industrial districts through the creative re-use of the many industrial brownfield sites to keep the city competitive in the new regional and global economies.
3 Focusing on stabilizing the city's viable neighborhoods.
4 Restoring the city's downtown core, which retains cultural assets such as parks, art museums, and a major university.

The plan's bold vision for smart decline drew national acclaim, winning the American Planning Association's prestigious National Planning Excellence Award for Public Outreach in 2007. It was also named one of *The New York Times*' "Best Ideas of the Year". Educating people about the importance of planning for a smaller city was a primary goal of the three-year planning and visioning process, and Youngstown residents were engaged in and accepted the idea of a smaller city. The concept gained popular acceptance such that other cities have followed suit with policies and plans based on rightsizing ideology, including Cleveland, Ohio (Box 7.2); Detroit and Flint, Michigan; and Rochester, New York (Hackworth 2015).

Some success, some challenges

In 2009, the Youngstown Neighborhood Development Corporation (YNDC) was created between the city and a regional foundation to assist in neighborhood-level initiatives and projects. Among its projects is the Iron Roots Urban Farm, a 1.7-acre farm and training center that acts as a fresh food source and training center to educate residents interested in growing their own food. The farm is located in the Idora neighborhood, one of the few neighborhoods to have a plan created as a result of *Youngstown 2010*. The Idora Neighborhood Association (INA) has numerous programs and events that engage residents in improving their neighborhood. The INA hosts a monthly Community Workday where residents work to improve vacant properties, including boarding up buildings, greening vacant lots, and preparing sites for projects.

However, the *Youngstown 2010* plan is not without its challenges, mainly that "the City cannot afford single handedly [to] do all that this plan calls for on its own" (City of Youngstown 2005). Although *Youngstown 2010* was innovative for presenting a new vision for shrinking cities and was successful in raising the city's profile and perception, the city has encountered challenges implementing many of the plan's goals due to a lack of fiscal capacity and political motivation after Mayor Jay Williams left to join President Barack Obama's cabinet. For example, fiscal limitations have forced the city to prioritize the most blighted areas for demolition and target the least expensive properties first, resulting in scattershot and unfocused demolition activity that does little to stabilize healthier neighborhoods and retain the existing population.

With limited funds, Youngstown has turned to the state and federal government for funding and regulatory assistance. In both cases, help has been limited. The fiscal climate in Ohio has left many plan elements predicated on unallocated state funds, which are likely to be unrealized (Hackworth 2015). Despite a high level of local funding for demolitions, costly regulations imposed by higher levels of government keep costs to demolish a building high. A waiver that would have lessened regulatory costs and allowed for increased demolitions was denied by the Environmental Protection Agency (EPA). In fact, subsequent regulations have actually *increased* the average cost of carrying out a building demolition in the city.

Box 7.2 Re-imagining vacant land in Cleveland

Cleveland, Ohio, is among America's worst shrinking cities, suffering sustained population loss since it peaked in the 1950 census at 914,808. Despite a decrease in the rate of decadal population loss from –23.6 percent in the 1970s to –5.4 percent in the 1990s, population loss accelerated in the 2000s, declining 17.1 percent by 2010. The only city to lose a greater percentage of population in the 2000s was Hurricane Katrina-ravaged New Orleans (Keating 2009). In total, Cleveland's population shrank from 914,808 in 1950 to 389,521 in 2014 (2014 ACS) – a total decrease of 57.4 percent.

With such extreme and sustained population decrease comes a reduction in the number of housing units and other buildings, as market conditions weaken to the point where structures transition from vacant to abandoned to ultimately demolished. In Cleveland, the number of housing units decreased from 264,100 in 1970 to 207,536 in 2010. The result was extensive numbers of vacant, municipally owned properties with no market demand. In 2008, there were approximately 20,000 vacant lots, the equivalent of 3,300 acres or 5.15 square miles of once-development properties (Kent State University 2008). By 2011, the amount increased to 3,750 acres, roughly 5.9 square miles. It is likely that the area of vacant land in Cleveland currently exceeds 6.2 square miles, approximately the size of Shaker Heights, a first-ring Cleveland suburb.

Envisioning vacant land as a community asset

With the backdrop of thousands of vacant parcels, *Re-Imagining a More Sustainable Cleveland* (*Re-Imagining Cleveland*) was initiated as a vacant land study project jointly undertaken by the City of Cleveland Planning Department, Kent State University, and Neighborhood Progress, Inc. The plan was ultimately adopted by the Cleveland City Planning Commission in 2008.

The plan seeks to create a new image of and for Cleveland, one that views vacant land not as a signal of distress and disorder but as an asset and an opportunity to rebuild and reimagine the city. Viewing land as an asset, the community pursues the reuse of land in order to "advance a larger, comprehensive sustainability strategy for the city, benefit low-income and underemployed residents, enhance the quality of neighborhood life, create prosperity in the city and help address climate change" (Kent State University 2008: 1).

The starting point for developing key vacant land use strategies was to seek out ways to benefit from the growing portfolio of land that, when possible, could support the city's redevelopment efforts. Additionally, re-use strategies centered on linking the natural and built environments in ways that improved, not detracted from, the quality of life for Clevelanders, including the ability to be food and energy self-reliant (Kent State University 2008)

The goals of the plan are based on a strategic decision-making matrix based on key factors:

1 To reuse vacant land in a manner that is productive and has a public benefit. Regardless of the end use, be it urban agriculture or redevelopment, it should offer an economic, environmental, and/or community benefit.
2 To improve local ecosystem function, from storm water management and soil restoration to improved wildlife habitat.
3 To eliminate the human health and environmental risks that contaminated vacant properties can impose.

A land-use decision-matrix provides a guide for evaluating each property based on economic, sustainability, and quality-of-life goals (Reichtell 2012). Strategies fall into three categories:

1 *Neighborhood Stabilization and Holding Strategies.* These strategies are aimed at properties that have the potential to be redeveloped within five years. These are low-maintenance and low-cost strategies that, while holding land for future development, also seek to stabilize neighborhoods by creating the appearance of stability and order, which produces new, positive images of a neighborhood. Strategies employed in these neighborhoods include:

   - Planting low-mow landscapes that require little maintenance but create the image of stability and stewardship.
   - Encouraging residential side-lot expansions and consolidation in neighborhoods whereby residents acquire neighboring properties and own and maintain them.
   - Planting trees in a strategic manner to still allow development in the future but improve the urban landscape in the interim.

2 *Green Infrastructure.* This strategy seeks to utilize vacant properties to expand and connect the city's existing green infrastructure while improving the function of ecological systems and remediating environmental contamination; increasing access to parks and recreational amenities; and improving public health. Strategy highlights include:

   - Utilizing properties with limited market demand to expand the city's parks and open space system. For example, properties that are adjacent to or near existing parks are prime candidates.
   - The legacy of heavy manufacturing has left many vacant sites requiring environmental remediation. Using alternative forms of remediation, such as bioremediation, that take a longer time to clean a site fit with the lack of market demand for these properties.

3 *Productive Landscapes.* The plan realizes that vacant sites can offer economically productive uses that support local residents, such as urban agriculture and the generation of renewable energy.

   - Urban agriculture offers the ability to help residents overcome a lack of access to fresh produce in neighborhoods lacking grocery stores, giving residents some level of food security.
   - Renewable energy can be generated on vacant properties by wind, solar, and geothermal energy, based on the site criteria of each technology.

This plan was developed by more than thirty non-profit and local government agencies that, among many goals, sought to assist and empower stakeholders, neighborhoods' residents, and community organizations in the battle against vacant property. The *Re-Imagining Cleveland Vacant Land Re-Use Pattern Book* (*Pattern Book*), a companion to *Re-Imagining Cleveland*, seeks to "provide inspiration, guidance and resources for community groups and individuals who want to create productive benefit from vacant land in their neighborhood" (Kent State University 2009: 1). The book is a guidebook for those interested in using vacant land in their neighborhoods to create new urban spaces that connect people and spaces together. The plan provides drawings for various vacant land treatments with cost estimates. Treatments include community gardens, rain gardens, trails, parks, pocket parks, and native planting plans.

After the plan was adopted by the City of Cleveland, Neighborhood Progress, Inc. funded and managed pilot projects in six Cleveland neighborhoods that resulted in the twenty small-lot projects. The success of these initial projects lead to $500,000 in additional funding from the Department of Housing and Urban Development's Neighborhood Stabilization Program (NSP) as well as funds from local foundations. With more than fifty projects in place, from small parks and walking paths to rain gardens and urban farms, *Re-Imaging Cleveland* has continued to connect people to the tools and funding to help them reclaim their neighborhoods and reimagine their city.

Concluding remarks

By addressing alternatives to the existing *pro-growth* approaches in American urban and regional policy discourses with the help of real-world examples of rightsizing policy instruments, this chapter made two subtle implications. First, there appears to be a sea change ahead, or even in progress, in the world of shrinking cities planning. In both research and practice, new rightsizing and smart decline conceptualizations are challenging, and in some cases supplanting, the growth-oriented paradigms of the past. Increasing attention is being paid to issues of optimal scale, social justice, and public participation in planning and decision-making processes, while the belief that shrinking cities will re-grow to their former scales is being abandoned/questioned. Yet, second, the shrinking places where planning approaches have ostensibly already been re-oriented toward rightsizing are still struggling to escape the downward spiral of shrinkage and decline. While such circumstances might call these rightsizing approaches and policies into question, it should be noted that a paradigm shift generally requires time, snowballing levels of commitment, and structural changes (Nadeau 2006).

These latter requirements – escalating commitment and attendant structural changes – broaden the scope of shrinkage and decline beyond an affected neighborhood or an affected city. Specifically, there is only so much that a place can accomplish on its own, with its own assets. In the United States, for instance, for a city to attempt *consolidation* or one of the other wide-reaching policies surveyed above, it would almost certainly need resources from regional, state, or federal agencies. However, if decision-makers in these related agencies still consider pro-growth to be the only viable option for urban development, then they might not provide the support necessary for the city to realize its consolidation goals. In other words, due to the linkages and relationships that exist in systems that produce patterns of *parasitic urbanization*, it is generally not efficacious for shrinking cities to go it alone in rightsizing efforts. Cooperation – between governments and stakeholders at various levels of the system – seems to be a necessary component of such efforts. Thus, to pave the way for broader patterns of commitment to rightsizing, it might be necessary to examine possibilities to make intergovernmental cooperation easier and more mutually beneficial. The next chapter explores these issues in greater detail.

Note

1 Oswalt (2006) uses the term *deconstructing* rather than *disassembling*. However, the latter is adopted here to avoid confusion with the rightsizing policy tool *deconstruction*, which involves tearing down buildings in a way that conserves some of their materials and components.

References

Ahn, T. K., and Elinor Ostrom. 2008. "Social Capital and Collective Action." In *The Handbook of Social Capital*, edited by Dario Castiglione, Jan W. Van Deth and Guglielmo Wolleb, 70–100. Oxford: Oxford University Press.

Axel-Lute, Miriam. 2007. "Small Is Beautiful – Again." *Shelterforce Online* (Summer). http://www.shelterforce.org/article/657/small_is_beautiful_again/

Bales, Kevin. 1985. "Determinants in the Perceptions of Visual Blight." *Human Ecology* 13 (3):371–387.

Batty, Michael. 2013. "A Theory of City Size." *Science* 340 (6139):1418–1419.

Beauregard, Robert A. 2012. "Growth and Depopulation in the United States." In *Rebuilding America's Legacy Cities: New Directions for the Industrial Heartland*, edited by Alan Mallach, 20–21. New York: The American Assembly.

Blomley, Nicholas K. 2004. *Unsettling the City: Urban Land and the Politics of Property*. New York, NY: Routledge.

Buffalo ReUse. http://www.buffaloreuse.org/.

City of Youngstown. 2005. *Youngstown 2010 Citywide Plan*. Youngstown, OH.

Cumming, Graeme S., David H.M. Cumming, Charles L. Redman. 2006. "Scale Mismatches in Social-Ecological Systems: Causes, Consequences, and Solutions." *Ecology and Society* 11 (1):14.

Daniels, Tom. 2001. "Smart Growth: A New American Approach to Regional Planning." *Planning Practice and Research* 16 (3–4):271–279.

Dewar, Margaret, and June Manning Thomas, eds. 2012. *The City after Abandonment*. Philadelphia: University of Pennsylvania Press.

Fischel, William A. 1987. *The Economics of Zoning Laws: A Property Rights Approach to American Land Use Controls*. Baltimore: Johns Hopkins University Press.

Frazier, Amy E., and Sharmistha Bagchi-Sen. 2015. "Developing Open Space Networks in Shrinking Cities." *Applied Geography* 59:1–9.

Frazier, Amy E., Sharmistha Bagchi-Sen, and Jason Knight. 2013. "The Spatio-Temporal Impacts of Demolition Land Use Policy and Crime in a Shrinking City." *Applied Geography* 41:55–64.

Ganning, Joanna P., and J. Rosie Tighe. 2014. "Assessing the Feasibility of Side Yard Programs as a Solution to Land Vacancy in U.S. Shrinking Cities." *Urban Affairs Review* 51 (5):708–725.

Glaeser, Edward L., and Joseph Gyourko. 2006. "Housing Dynamics." In *Working Paper Series: National Bureau of Economic Research*. NBER Working Paper 12787, National Bureau of Economic Research, Cambridge, MA.

Hackworth, Jason. 2014. "The Limits to Market-Based Strategies for Addressing Land Abandonment in Shrinking American Cities." *Progress in Planning* 90:1–37.

Hackworth, Jason. 2015. "Rightsizing as Spatial Austerity in the American Rust Belt." *Environment and Planning A* 47 (4):766–782.

Hollander, Justin B., and Bernard Cahill. 2011. "Confronting Population Decline in the Buffalo, New York, Region: A Close Reading of the 'Erie-Niagara Framework for Regional Growth'." *Journal of Architectural and Planning Research* 28 (3):252–267.

Hollander, Justin B., and Jeremy Németh. 2011. "The Bounds of Smart Decline: A Foundational Theory for Planning Shrinking Cities." *Housing Policy Debate* 21 (3):349–367.

Hollander, J.B., Karina Pallagst, Terry Schwartz, and Frank J. Popper. 2009. "Planning Shrinking Cities." *Progress in Planning* 72 (4):223–232.

Hospers, Gert-Jan. 2014. "Urban Shrinkage in the EU." In *Shrinking Cities: A Global Perspective*, edited by Harry W. Richardson and Chang Woon Nam, 47–58. London: Routledge.

Hummel, Daniel. 2015. "Right-Sizing Cities in the United States: Defining Its Strategies." *Journal of Urban Affairs* 37 (4):397–409.

Kantor, Paul. 2010. "City Futures: Politics, Economic Crisis, and the American Model of Urban Development." *Urban Research & Practice* 3 (1):11.

Keating, W. Dennis. 2009. "Redevelopment of Vacant Land in the Blighted Neighborhoods of Cleveland, Ohio Resulting from the Housing Foreclosure Crisis." ISA RC 43 Conference – Housing Assets, Housing People, Glasgow, Scotland.

Kent State University. 2008. *Re-Imagining a More Sustainable Cleveland*. Cleveland: Cleveland Urban Design Collaborative, Kent State University.

Kent State University. 2009. *Re-Imagining Cleveland Vacant Land Re-Use Pattern Book*. Cleveland: Cleveland Urban Design Collaborative, Kent State University.

Knox, Paul L. 2011. *Cities and Design, Critical Introductions to Urbanism and the City*. New York: Routledge.

Kuhn, Brad. 2011. "What Is Downzoning and When Is It Compensable?" Nossaman LLP. http://www.californiaeminentdomainreport.com/2011/11/articles/inverse-condemnation-regulatory-takings/what-is-downzoning-and-when-is-it-compensable/.

LaCroix, Catherine J. 2011. "Urban Green Uses: The New Renewal." *Planning & Environmental Law* 63 (5):3–13.

Lawson, Laura, and Abbilyn Miller. 2012. "Community Gardens and Urban Agriculture as Antithesis to Abandonment – Exploring a Citizenship-Land Model." In *The City after Abandonment*, edited by Margaret Dewar and June Manning Thomas, 400. Philadelphia: University of Pennsylvania Press.

Leo, Christopher, and Kathryn Anderson. 2006. "Being Realistic about Urban Growth." *Journal of Urban Affairs* 28 (2):169–189.

Linkon, Sherry Lee, and John Russo. 2002. *Steeltown USA: Work and Memory in Youngstown*. Lawrence: University Press of Kansas.

Madanipour, Ali. 2006. "Roles and Challenges of Urban Design." *Journal of Urban Design* 11 (2):173–193.

Mahoning Valley Organizing Collaborative. n.d. "Fighting Disinvestment." http://www.mvorganizing.org/index.php?option=com_content&view=category&layout=blog&id=60&Itemid=541.

Mallach, Alan. 2010. *Bringing Buildings Back: From Abandoned Properties to Community Assets: A Guidebook for Policymakers and Practitioners*. 2nd ed. Montclair, NJ: National Housing Institute.

Mallach, Alan. 2011. "Comment on Hollander's 'the Bounds of Smart Decline: A Foundational Theory for Planning Shrinking Cities'." *Housing Policy Debate* 21 (3):369–375.

Mallach, Alan. 2012. "Laying the Groundwork for Change: Demolition, Urban Strategy, and Policy Reform." Brookings Metropolitan Policy Program, Washington, DC, The Brookings Institution. http://www.brookings.edu/~/media/research/files/papers/2012/9/24-land-use-demolition-mallach/24-land-use-demolition-mallach.pdf

Marshall, Stephen. 2009. *Cities Design & Evolution*. London, UK: Routledge.

McGuirk, P.M. 2001. "Situating Communicative Planning Theory: Context, Power, and Knowledge." *Environment and Planning A* 33 (2):195–217.

Nadeau, Robert. 2006. *The Environmental Endgame: Mainstream Economics, Ecological Disaster, and Human Survival*. New Brunswick, NJ: Rutgers University Press.

Nassauer, Joan Iverson, and Julia Raskin. 2014. "Urban Vacancy and Land Use Legacies: A Frontier for Urban Ecological Research, Design, and Planning." *Landscape and Urban Planning* 125:245–253.

National Vacant Properties Campaign. 2005. "Vacant Properties: The True Costs to Communities." National Vacant Properties Campaign. http://www.smartgrowthamerica.org/research/policy-analysis-vacant-properties/.

Olavsrud, Thor. 2013. "Big Data Drives City of Buffalo's Operation Clean Sweep." http://www.cio.com/article/2382750/government/big-data-drives-city-of-buffalo-s-operation-clean-sweep.html.

Ostrom, Elinor. 1990. *Governing the Commons: The Evolution of Institutions for Collective Action*. Cambridge: Cambridge University Press.

Ostrom, Elinor. 2005. *Understanding Institutional Diversity*. Princeton: Princeton University Press.

Ostrom, Elinor. 2009. "Design Principles of Robust Property Rights Institutions: What Have We Learned?" In *Property Rights and Land Policies*, edited by Gregory K. Ingram and Yu-Hung Hong, 25–51. Cambridge: Lincoln Land Institute.

Ostrom, Vincent. 1972. "Theory of Public Policy." *Policy Studies Journal* 1 (1):11–14.

Oswalt, Philipp. 2006. *Shrinking Cities, Volume 2 — Interventions*. Ostfildern-Ruit: Hatje Cantz Verlag.

Popper, Deborah E., and Frank J. Popper. 2002. "Small Can Be Beautiful." *Planning* 68 (7):20–23.

Register, Richard. 2006. *Ecocities: Rebuilding Cities in Balance with Nature*. Gabriola Island, Canada: New Society Publishers.

Reichtell, Bobbi. 2012. "Case Study: Reimagining Cleveland: Pilot Land Use Projects." In *Rebuilding America's Legacy Cities: New Directions for the Industrial Heartland*, edited by Alan Mallach, 185–188. New York: The American Assembly.

Rhodes, James, and John Russo. 2013. "Shrinking 'Smart'?: Urban Redevelopment and Shrinkage in Youngstown, Ohio." *Urban Geography* 34 (3):305–326.

Ryan, Brent D. 2012. *Design after Decline: How America Rebuilds Shrinking Cities*. Philadelphia, PA: University of Pennsylvania Press.

Sager, Tore. 2011. "Neo-Liberal Urban Planning Policies: A Literature Survey 1990–2010." *Progress in Planning* 76:147–199.

Savitch, H.V. 2011. "A Strategy for Neighborhood Decline and Regrowth: Forging the French Connection." *Urban Affairs Review* 47 (6):800–837.

Schatz, Roland. 2012. "Overcoming the Risk of Stereotypes: How Strategic Communications Can Facilitate Sustainable Place Branding." In *International Place Branding Yearbook 2012: Managing Smart Growth and Sustainability*, edited by Frank M. Go and Robert Govers, 147. New York: Palgrave Macmillan.

Schilling, Joseph, and Jonathan Logan. 2008. "Greening the Rust Belt: A Green Infrastructure Model for Right Sizing America's Shrinking Cities." *Journal of the American Planning Association* 74 (4):451–466.

Schilling, Joseph M., and Alan Mallach. 2012. *Cities in Transition: A Guide for Practicing Planners*. Vol. 568, *Planning Advisory Service*. Chicago: American Planning Association.

Temkin, Kenneth, and William M. Rohe. 1998. "Social Capital and Neighborhood Stability: An Empirical Investigation." *Housing Policy Debate* 9 (1):61–88.

Tumber, Catherine. 2012. *Small, Gritty, and Green: The Promise of America's Smaller Industrial Cities in a Low Carbon World*. Cambridge: The MIT Press.

Turchin, Peter. 2003. *Complex Population Dynamics: A Theoretical/Empirical Synthesis*. Vol. 35, *Monographs in Population Biology*. Princeton: Princeton University Press.

United States Environmental Protection Agency. n.d. "Brownfield Overview and Definitions." http://www.epa.gov/brownfields/brownfield-overview-and-definition

Wagenaar, Hendrik. 2007. "Governance, Complexity, and Democratic Participation: How Citizens and Public Officials Harness the Complexities of Neighborhood Decline." *The American Review of Public Administration* 37 (1):17–50.

Warner, Daniel M. 2006. "Commentary: 'Post-Growthism': From Smart Growth to Sustainable Development." *Environmental Practice* 8 (3):169–179.

Weaver, Russell. 2015. "A Cross-Level Exploratory Analysis of 'Neighborhood Effects' on Urban Behavior: An Evolutionary Perspective." *Social Sciences* 4 (4):1046–1066.

Weaver, Russell, and Chris Holtkamp. 2015. "Geographical Approaches to Understanding Urban Decline: From Evolutionary Theory to Political Economy . . . and Back?" *Geography Compass* 9 (5):286–302.

Weaver, Russell, and Jason Knight. 2015. "Demonstrative Urban Policy: Evidence from a Pilot Study of Neighborhood 'Clean Sweeps' in Buffalo, NY." Annual Meeting of the Association of American Geographers, Chicago.

Weaver, Russell C. 2013. "Re-Framing the Urban Blight Problem with Trans-Disciplinary Insights from Ecological Economics." *Ecological Economics* 90:168–176.

Weaver, Russell C., and Sharmistha Bagchi-Sen. 2013. "Spatial Analysis of Urban Decline: The Geography of Blight." *Applied Geography* 40:61–70.

Wilson, David Sloan. 2014. *Does Altruism Exist?: Culture, Genes, and the Welfare of Others*. New Haven: Yale University Press.

Wilson, David Sloan, Elinor Ostrom, and Michael E. Cox. 2013. "Generalizing the Core Design Principles for the Efficacy of Groups." *Journal of Economic Behavior & Organization* 90 (Supplement):S21–S32.

8 Challenges and prospects of regional governance

Chapter 7 concluded by speculating that *cooperation* – between government and non-government stakeholders at all levels, from households and neighborhoods to local, state, and federal institutions – is vital to any effort aimed at mitigating or eliminating the harmful consequences of shrinkage and decline. The current chapter describes existing U.S. government and governance systems in some detail and then explores ways in which greater cooperation among the many actors that are affected by shrinkage and decline can be facilitated.

First, consider the following takeaways from Chapters 5, 6, and 7:

1 The processes and outcomes of shrinkage and decline are not always confined to central city borders (Ch. 5).
2 Consistent with the *American model of urban development*, a stereotypical U.S. metropolitan region is a fragmented collection of independent, decision-making authorities (i.e., *general purpose local governments* [see definition below]) that compete with one another over various forms of capital (Ch. 6).
3 Planning scholars and practitioners have proposed a variety of innovative *rightsizing* policy instruments for improving conditions in shrinking cities; however, the ability of a city to enact many of these policies is often undermined by insufficient local finances and/or lack of state- or federal-level support (Ch. 7).

Together, these three points describe a tension and an apparent paradox. The tension, stated in the second bullet, is inter-municipal rivalry. The conventional view of the American model of urban development is that municipal governments are self-interested institutions wishing to maximize their economic position relative to other cities (Leo and Anderson 2006). This tension creates a paradox. Namely, to the extent that (i) shrinking cities lack the internal capacity to address their attendant problems, and (ii) negative causes and consequences of shrinkage extend beyond central city limits, inter-municipal cooperation necessarily emerges as a public policy priority in regions affected by shrinkage (Kübler, Scheuss, and Rochat 2012). This observation constitutes a paradox precisely because the American model tends to portray competition – the model's mechanism of urban development – as being incompatible with, and mutually exclusive of, *cooperation*. Recall that competition produces "success" in this worldview, leaving no room for cooperation (Kantor 2010; see Ch. 6).

Responding to this apparent paradox (i.e., the disconnect between the prevailing model of inter-municipal competition on one hand and the necessity of inter-municipal cooperation on the other), students of shrinking cities have embraced a growing **new regionalism** movement that advocates for coordinated regional management of America's urban problems (Wheeler 2002). The movement recommends changing the structure of American metropolitan government to increase the likelihood of inter-governmental cooperation. However, at least two different perspectives on regional coordination have come out of the new regionalism movement. The first, sometimes called the neoprogressive view (Feiock 2007), argues that because inter-municipal competition presently dominates the urban development landscape, cooperation is best achieved by consolidating existing governmental units into a centralized regional government (Lowery 2000). In other words, this view calls for "de-fragmenting" the urban metropolis, in order to align the interests of competing groups by placing them inside a uniform political/jurisdictional context. Thus, centralization is thought to make communities, which were formerly fully autonomous, interdependent, and mutually staked in the overall success of the region (Rusk 2013).

By contrast, the second perspective argues that regional cooperation can and does emerge from interactions within existing, decentralized institutional networks (Feiock 2013). More precisely, contrary to conventional framings of fragmented governance, competition is not the sole mechanism of urban development in the real world. Competition and cooperation coexist in American regions (Feiock 2004). Moreover, where inter-municipal cooperation among fragmented institutions is possible, decentralized governance might be more effective than centralized regional governments at solving urban problems (Grassmueck and Shields 2010; Lee et al. 2011). In this sense, the apparent paradox is not a paradox at all the American model of urban development, and its singular focus on competition, simply does not reveal the whole picture.

This debate over the appropriate mechanism(s) for achieving regional coordination is of critical importance to shrinking cities research and policy communities. Nonetheless, it has not found its way into most books on the topic. This is not to say the debate is absent from the shrinking cities literature (Rybczynski and Linneman 1999). However, considerations of its origins in collective action theory and urban politics are relatively sparse. Accordingly, the next section briefly engages with these origins to describe the fundamental *institutional collective action problem* that local governments and other stakeholder groups in and around shrinking cities will need to address if they are to achieve some degree of regional cooperation. This discussion leads to a distinction between *government*, which describes the formal metropolitan organizations of government (e.g., *general purpose* and *special purpose* local governments), and *governance*, which extends beyond formal institutions to include informal actors and interconnections that contribute to public decision-making in regions that contain shrinking cities (Savitch and Vogel 2000).

Notably, despite common black-and-white perceptions that regionalism is attainable only through consolidation or decentralization, bridging the discussions of government organizations and governance structures reveals that the two debated solutions are better viewed as points on a continuum rather than parts of a dichotomy. That is, there are several general regional governance mechanisms that might facilitate inter-municipal cooperation in shrinking city metropolitan regions. The final sections of this chapter describe some of these mechanisms with the help of real-world examples from shrinking cities.

The institutional collective action problem

In general, a **collective action problem** occurs when (1) the decisions and outcomes of multiple actors are correlated, such that (2) decision-makers can only achieve higher-valued outcomes [relative to the status quo] by cooperating with each other, and (3) where it is difficult to exclude non-cooperative actors from receiving benefits created by cooperative actors (Ostrom 2008). A classic example of a collective action problem is the *tragedy of the commons* (Hardin 1968). Consider a resource that is not legally or functionally controlled by a single decision-maker and can be openly accessed by multiple individuals. Within a shrinking city, such a "resource" might take the form of an abandoned vacant lot to which entry is neither closely monitored nor actively deterred. Local community members who live near such a lot presumably have an incentive to use it – given that it is abandoned and unmonitored – for their own benefits. In certain neighborhoods in Detroit, for example, residents sometimes use vacant lots as locations for (illegally) disposing of unwanted items (Dewar and Thomas 2012). Even though these ad hoc dump sites are blighting factors that devalue nearby properties, contribute to negative neighborhood images, and reduce local quality of life, all residents face the same incentive to use the lot for their benefit (e.g., to discard trash).

Under this arrangement, any resident who does not make use of the vacant site bears the cost of others' dumping activity in the form of lower neighborhood quality of life; but without the benefit of dumping. Hence, while all residents would be collectively better off if no one illegally dumped at the abandoned site, from each individual resident's perspective, it is not rational to forego illegal dumping when others in the neighborhood adopt that behavior. The result is a **tragedy of the commons**, or an overexploitation of the open access resource. The only way to overcome the tragedy is with a mechanism that facilitates cooperation between residents (that is, through *collective action*). One possible coordination mechanism for achieving collective action in the vacant lot example is a cooperative ownership model that extends owner rights over the abandoned lot jointly to all members of the neighborhood (see Ch. 7).

Just as individual decision-makers face collective action problems such as the foregoing vacant lot dilemma, institutional decision-makers such as local governments routinely make choices that affect one another. For example, downzoning to permit agricultural land use at the edge of a shrinking city (see Ch. 7) inexorably affects adjacent jurisdictions. Agricultural activities generate odors and sounds that might be regarded as nuisances by landowners across the city's borders. More harmfully, pesticide drift could lower the environmental quality of neighboring communities, among other problems. While this particular example is probably a rare occurrence in the current landscape of shrinking cities, meant only to illustrate the nature of the types of *institutional* collective action problems faced by metropolitan governments, subtler forms of uncoordinated land use and other local decision-making activities are ubiquitous features of fragmented political landscapes (Kwon and Park 2014).

For instance, a 2015 report by the Organization for Economic Co-operation and Development (OECD) on metropolitan governance laments that "there are numerous cities where certain transport modes – for no apparent economic reason – end at administrative borders" (OECD 2015: 52). Since functional borders (e.g., commuting regions) extend beyond administrative borders, this type of disconnected service provision results in substantive regional welfare losses, such as longer commute times and more traffic congestion for residents who must travel between jurisdictions. More pointedly, many of the more than ninety municipalities in St. Louis County, Missouri (part of the metropolitan region that contains the shrinking city of St. Louis, Missouri), have taken to using excessive "speed traps, traffic tickets, and petty fines" to fill their dwindling city coffers (Badger 2015). While the potential exists for these municipalities to combine efforts and share, for instance, fire and police protection resources, they continue to approach institutional collective problems as if their decisions and positions are independent of one another. Why, then, do municipalities that stand to gain from collective action fail to cooperate?

Generally speaking, inter-municipal cooperation requires that "benefits to potential collaborators are high, and the *transaction costs* of negotiating, monitoring, and enforcing agreements are low" (Feiock 2007: 51; emphasis added). The first of these requirements suggests that the likelihood of regional cooperation depends on the size of the anticipated benefits from working together. It is perhaps for this reason that suburban municipalities are often reluctant to contribute resources to improving conditions in shrinking central cities in their regions – say, by sharing their tax revenues (e.g., Orfield 1997). Plainly, relatively healthy and wealthy suburbs are unlikely to expect their returns on these investments to outweigh the costs (Favro 2010). Such a perception is plausibly due to the absence of the second requirement for institutional collective action: low transaction costs.

Transaction costs refer to the set of impediments that foreclose on opportunities for decision-makers to form mutually beneficial alliances. With respect to city-suburb revenue sharing, one relevant type of transaction cost stems from *incomplete information*. If a suburb has a successful economy, decision-makers (especially voters) in that community might lack motivation to transfer portions of their hard-earned wealth to a declining central city. However, this outlook is decidedly myopic, in that it fails to consider how continued central city shrinkage and decline might spill over to the suburb over time (e.g., Fig. 5.5). Another important source of transaction costs are the state-level rules that influence the types of cooperative arrangements and regional solutions that are available to local governments. These issues are grappled with in more detail later in the chapter (see especially Boxes 8.1–8.3).

For now, we conclude this section by pointing out that the requirements just discussed – high/perceptible collective benefits and low transaction costs – are necessary for the emergence of *voluntary* regional cooperation. The absence of one or both of these conditions does not constitute an outright problem for regional cooperation. In contrast, coercive or incentivizing tools might be deployed by states to require or facilitate (through reducing transaction costs) inter-municipal cooperation. The state of Florida, for instance, requires local governments to participate in regional councils that promote consistency in intra-regional comprehensive planning efforts (Kwon and Park 2014). While coercive means can be used to mandate regional cooperation, research has found that such means can crowd out voluntary forms of cooperation (Ostrom 2005; Kwon and Park 2014), and decrease the ability of decision-makers to reach consensuses on key issues (Feiock, Tao, and Johnson 2004; Post 2004). For these reasons, coercive or incentivizing mechanisms might be best applied in situations where potential benefits from cooperation and transaction costs are *both* high. With respect to the former opportunity, the prospect of high benefits makes regional cooperation a worthwhile pursuit; however, the latter constraint – high transaction costs – implies that institutions will not voluntarily enter into regional partnerships.

Government and governance

The preceding section described the nature of the institutional *collective action problem* that decision-makers in regions with shrinking cities must overcome if they are to achieve some degree of regional cooperation. Crucially, decision-makers are left to find ways to identify the costs and benefits relevant to a particular proposal for collective action, and all parties potentially affected by that proposal need to contribute to the discourse and negotiate the eventual outcome. The participation of impacted parties implies that *transaction costs* to participate in the discourse are low, such that parties are in fact able to come together and form mutually beneficial alliances. On that foundation, a fundamental question that has not yet been addressed is: who are these "decision-makers" and "parties" that might or might not join together in regional cooperation?

To answer this question, it is necessary to distinguish between govern*ment* and govern*ance*. **Government** consists of the formal institutions, elections, and administrative structures that make binding public decisions on behalf of persons and entities located within a precisely defined set of spatial boundaries (Savitch and Vogel 2000). These formal institutions are characterized by *legitimacy* – i.e., they possess constitutional or legislative authority to make decisions on behalf of the persons and entities within their governing boundaries – and *accountability* – i.e., they are required to reveal, explain, and justify their actions to the publics for whom they are acting as decision-making agents (Morgan and Yeung 2007).

Governance refers to the broader process by which diverse stakeholders make trade-offs between competing interests, in shared environments, for the purpose of protecting and enhancing the public realm (Oakerson 2004). Governance thus extends beyond formal institutions to create a large and heterogeneous constellation of government and nongovernment decision-makers – e.g., municipal governments, voters in referenda, grant-making institutions, businesses, civic groups, and community and neighborhood associations, among others. Unlike formal institutions of *government*, informal institutions of *governance* are unlikely to have legitimacy and accountability. Consequently, such institutions tend not to make binding and authoritative public decisions. Instead, they regularly leverage their organizational capacities to influence and shape the public decisions made by formal institutions. In that sense, they are pivotal "players" in the institutional collective action "games" acted out in metropolitan regions (Feiock 2013).

With the government-governance distinction in place, the next subsection defines the formal *government* organizations and structures found in American shrinking cities. The arrangements, powers, and functions of these organizations necessarily affect the types of *governance* structures that emerge in metropolitan regions. These governance structures, and their ties to the centralization-decentralization debate from the new regionalism movement introduced earlier in this chapter, are examined in the final part of this section.

The governmental framework of U.S. metropolitan regions

The U.S. federalist system of government guarantees that geographic territories are under the control of two layers of government. The higher, *national* or *federal* level of government has legitimacy for, authority over, and is accountable to all persons and entities within the United States and its territories. The U.S. Constitution endows the federal government with an explicit set of powers and responsibilities. In general, these powers deal with matters of (inter)national concern. According to the Tenth Amendment, the second, *state* level of government is endowed with all governmental powers that are not specifically (1) granted to the national government in the Constitution or (2) proscribed by the Constitution. These state powers typically deal with matters of domestic as opposed to international affairs. In addition, state governments only have legitimacy for, authority over, and accountability to the persons and entities within their individual state borders. Furthermore, because different states exercise and interpret their powers differently, each of the fifty state governments in the United States has its own set of rules and regulations, which generates considerable variation in the landscape of U.S. *government*.

Adding to this variation, all state governments in the United States formally vest portions of their public decision-making authority in *local* units of government. These local governments can be placed into two broad categories. First, **general-purpose local governments** are (1) *organized entities* that possess (2) *governmental character*, (3) *substantial autonomy*, and (4) *provide a broad array of public services and perform a variety of functions* (U.S. Census Bureau 2013). Existence as an "organized entity" means that a general purpose government holds some corporate *powers*, including but not limited to the following (U.S. Census Bureau 2013; Platt 2014):

- have a name;
- sue and be sued;
- enter into contracts;
- acquire and dispose of property;
- levy taxes;
- regulate citizen conduct through local ordinances; and
- borrow against future revenues.

Second, in addition to existing as corporate organizations, general purpose local governments serve, have legitimacy for, and are accountable to the public inside their sub-state geographic boundaries. This *governmental character* exists when governing boards or officers are either popularly elected or appointed to their positions by popularly elected officials. Third, general purpose governments are said to possess *substantial autonomy* if they have considerable fiscal and administrative independence. **Fiscal independence** means that the government has the power to determine its own budget, tax levies, service fees, and debt issuances. **Administrative independence** means that the government performs functions that are different from its creating (state) government. Finally, general purpose governments, as their name suggests, provide a broad array of services and perform a variety of functions, including but not limited to:

- police and fire protection;
- sewer and water provision;
- transportation infrastructure and maintenance;
- public transportation;
- social services;
- public libraries;
- solid waste collection and disposal;
- parks and recreation; and
- court and elections administration.

The two main types of general purpose local government in U.S. metropolitan regions are **counties** and **municipalities** (including **townships**). Important particulars of these forms of government are discussed in Boxes 8.1 and 8.2.

Box 8.1 County and municipal forms of government

The following descriptions of *counties*, *municipalities*, and *townships* come from the National League of Cities (NLC), an advocacy organization with a mission of building the capacity of general purpose local governments and their leaders.

Counties

Early state constitutions originally created counties to serve as the administrative arms of state government, performing state-mandated duties, including property assessment and record keeping. Historically, counties were established without the consent of the voters, possessed no charter or legislative powers, performed no business or proprietary functions, and shared immunity with the state from suit. As populations grew and suburbs formed across the nation post-World War I, the role of local government was strengthened. After World War II, the urban populations began to spread beyond city boundaries into the suburbs, and county governments were increasingly called upon to provide services such as child welfare and consumer protection. County governments began to receive greater autonomy from the states, generate increasing revenues, and accept stronger political accountability. As a result, the number of counties that have their own charter has grown tremendously to 3,033 in 2007. Organized county governments are found in every state except Connecticut and Rhode Island – which have geographic regions called counties but without functioning county governments – and the District of Columbia. Counties are known as boroughs in Alaska and parishes in Louisiana. There are also limited portions of other states in which certain county areas lack a distinct county government. Priorities and service delivery responsibilities vary considerably among counties, as does their size and number. In general, counties have more mandates, less discretionary funds, and are more vulnerable to state budgetary action.

Municipalities

As of 2007, there are 19,492 municipal governments across the fifty states, but they vary widely according to *quantity* (Hawaii and the District of Columbia each have 1, Illinois has 1,299), *designation* (they may be called cities, towns, boroughs, districts, plantations, and villages), and *incorporation requirements* (Florida requires 1.5 persons per acre). Despite these variations, municipalities generally have similar powers and perform similar functions. Geographically, municipalities lie within counties, although they may cross county boundaries. Historically, towns and cities were distinguished by their distinct methods of deliberation. For example, all qualified citizens in a *town* deliberate and vote together, while *cities* have representatives who vote. Today, the distinction between towns and cities, and similarly with the other nomenclature, is one of population size.

Townships

Township governments are established to govern areas without a minimum population concentration. The 2007 Census of Governments counted approximately 3,000 fewer townships than municipalities across only twenty states, including New York, Maine, Illinois, and Kansas. Within these twenty states, there are different kinds of townships: the municipal or the civil, the incorporated or unincorporated, and the school, the judicial, and congressional.

Town government in its classic form is distinguished from township government, as the former is governed by an annual town meeting. Townships, if similar to municipalities, have a municipal form of government. Otherwise, townships are commonly governed by an elected board of three to five part-time trustees and rely almost exclusively on property taxes for revenue. Townships in New England, New Jersey, and Pennsylvania, for example, enjoy broad authority and perform functions similar to municipalities. Some New England townships govern schools, and Midwestern townships typically perform limited government functions."

Source (as quoted): http://www.nlc.org/build-skills-and-networks/resources/cities-101/city-structures/local-us-governments

Box 8.2 Dillon's Rule versus home rule: does it matter?

Because the U.S. Constitution omits any mention of local governments, such governments are understood to be "creatures . . . of the state, for the purpose of exercising a part of its power" (*Atkin v. Kansas* 1903). As a consequence, local government powers are derived directly from state constitutions, statutes, and/or charters (Richardson 2011). Many states where specific local government powers are enumerated in state laws are often referred to as Dillon's Rule states. **Dillon's Rule** is a rule of statutory construction that comes from an 1865 court case in which Judge John F. Dillon opined that local governments possess "only such powers as are specifically delegated to them by state law", and no others (Richardson, Zimmerman Gough, and Puentes 2003). That is, for a municipality to take an action that is not explicitly set forth in its state's laws, it would need to seek formal state approval. For that reason, some argue that Dillon's Rule inhibits local creativity, and, especially, stands in the way of regional cooperation (Frug 1984).

In contrast to Dillon's Rule, **home rule** is often used to describe a model similar to the federal-state relationship articulated in the U.S. Constitution, whereby states retain all powers that are not expressly denied to them. In other words, stakeholders regularly conflate *home rule* with a legal situation in which local governments "possess authority to act unless the state legislature particularly denies the authority" (Richardson 2011: 670). From this perspective, public officials in so-called Dillon's Rule states regularly "yearn for greater home rule authority", as they feel home rule would give them more freedom and possibilities with which to manage local and regional problems (Richardson, Zimmerman Gough, and Puentes 2003).

As it turns out, however, no type of actually existing home rule gives municipalities this degree of power. For the most part, home rule takes the form of a state constitutional provision that grant municipalities wide latitude to manage their "local affairs" without the state's direct oversight; but home rule "rarely provides substantial autonomy and freedom from state interference" (Richardson 2011: 671). Indeed, research has found there to be no substantive differences in the ability of municipalities in home rule and Dillon's Rule states to engage in regional cooperation (e.g., Richardson, Zimmerman Gough, and Puentes 2003); and some scholars have even found that local governments in Dillon's Rule states can hold as much, and occasionally more, power as their counterparts in home rule states (Bluestein 2006).

Coexisting with county and municipal general purpose local governments (see Box 8.1) are **special purpose local governments**. Broadly speaking, a special purpose government is one that performs either a single function or a small cluster of related functions for a specified area. Special purpose governmental units come in at least three varieties. A **special district**, such as a fire protection district, provides specialized services to the public within a fixed set of boundaries. These boundaries need not follow political administrative boundaries. Rather, special districts are overlaid onto targeted locations in metropolitan areas, and, for that reason, they are occasionally referred to as "institutional overlays" (Oakerson 2004). Special districts may be endowed with the power to levy taxes or fees.

A second type of special purpose local government is an **authority**. Authorities, such as organizations that provide and maintain low-income public housing facilities in urban areas, generally do not have the power to tax or levy fees. Further, they may or may not cut across municipal boundaries (Platt 2014). A third category of special purpose governments is the **school district**, which may be either an independent government unit or a dependent agency of state or local governments.

The first two panels in Figure 8.1 (a and b) map the total number of all general and special purpose local governments in the conterminous United States as of 2012, respectively, per square mile of land area, by metropolitan area. The data come from the periodic Census of Governments conducted by the U.S. Census Bureau.[1] Panels (c) and (d) in Figure 8.1 depict analogous maps for only those metropolitan areas that lie at least partially in states that feature one or more of the thirty-one "shrinking cities" from Table 2.3. Observe that, while metropolitan regions in Northeastern and Midwestern states (i.e., in the Rust Belt; see Beauregard 2009) appear to have more general purpose local governments per square mile than their counterparts elsewhere in the country (Fig. 8.1[a]); there is no striking

Figure 8.1 Distributions of (a) general purpose government density in metro regions, (b) special purpose government density in metro regions, (c) general purpose government density in states with shrinking cities, by metro region, and (d) special purpose government density in states with shrinking cities, by metro region

visual evidence to suggest that metropolitan regions containing cities that shrank over the past four decades are any more "fragmented" than other metro areas in their states (Fig. 8.1[c]). Similar statements can be made regarding the distribution of special purpose governments (Fig. 8.1[d]).

However, relationships detected with eyeball estimations via map comparisons tend to be unreliable. For that reason, Table 8.1 presents the results from nonparametric Wilcoxon signed-rank tests for differences in the medians (roughly) of four variables – (1) general purpose governments per square mile of land area; (2) general purpose governments per 1,000 persons; (3) special purpose governments per square mile of land area; and (4) special purpose governments per 1,000 persons – between metropolitan areas that contain one or more of the shrinking cities from Table 2.3, and those that do not [but are at least partially located in the same states as shrinking cities (see Fig. 8.1)].

The results are quite mixed. In the first place, metropolitan areas with shrinking cities seem to contain statistically significantly more general purpose governments

Table 8.1 Wilcoxon signed-rank tests for equality of medians in independent samples

Variable	*Shrinking City Present*	*Shrinking City Not Present*	*Difference*
General purpose governments per square mile of land area	0.05	0.04	0.01*
General purpose governments per 1,000 persons (2012 population)	0.12	0.15	–0.03
Special purpose governments per square mile of land area	0.02	0.02	0.0
Special purpose governments per 1,000 persons (2012 population)	0.06	0.10	–0.04*
n	21	91	

* $p < 0.05$; Note: the same patterns of results hold in t-tests of mean differences

per land area relative to metro areas in their same states that do not contain cities that shrank (severely) over the past four decades. Given that the median land area of metropolitan regions included in the analysis is 1,240.345 square miles, the difference observed in the first row of Table 8.1 suggests that a typical metro area with a shrinking city contains around twelve more general purpose local governments than a typical metro area without a shrinking city ($0.05 * 1240.345 \cong 62$; $0.04 * 1240.345 \cong 50$). This finding supports the claim that metropolitan regions with shrinking cities might be somewhat more fragmented with *independent municipalities* than regions that do not contain shrinking cities (Mallach and Brachman 2013). However, this apparent difference in areal government fragmentation does not extend to special purpose local governments (third row of Table 8.1). Moreover, when normalized by population instead of area, the median number of general purpose local governments in metro regions with shrinking cities is no different from the median number in regions without shrinking cities; and the median number of special purpose local governments is in fact *higher* for regions without shrinking cities compared to those with cities that shrank from 1970 to 2010 (Table 8.1).

Local government annexation and municipal (in)elasticity

One potential explanation for the higher number of general purpose governments *per unit area* in shrinking cities might relate to historical and contemporary patterns of annexation. **Annexation** refers to a process of municipal border expansion. In other words, the boundaries of local governments are subject to change. Not all territory in the United States lies within an *organized municipal entity*. These *unincorporated territories* generally receive local public services from their county governments because they do not have governing boards with power to levy taxes or perform general government functions. Frequently, it is economically more efficient for an adjacent general purpose municipal government, rather than a centralized county government, to provide these services. Under such circumstance, the general purpose municipal government might wish to *annex* the unincorporated territory. Annexed territory adds to the land area and population of the annexing municipality, and residents in annexed spaces become taxpaying citizens of the annexing government. Thus, annexation has the capacity to produce mutual benefits.

Cities that can expand their municipal boundaries via annexation, and/or those that contain undeveloped land capable of absorbing internal growth, are considered to be "elastic" (Rusk 2013). Cities that cannot expand their boundaries because they are not adjacent to unincorporated communities are considered "inelastic". Already incorporated territory (i.e., an existing general purpose government) is not annexable by legal standards. This stipulation brought about an annexation-incorporation race in the early and mid-twentieth century that contributed to the fragmented government patterns visible in U.S. metropolitan regions today. Namely, in the early twentieth century, as large cities aggressively annexed land adjacent to their borders to expand their tax bases, residents in growing unincorporated (suburban) communities were faced with a decision to incorporate or be annexed. Many of these communities opted for the former alternative, leading to an explosion of general purpose local governments in metropolitan regions (Fig. 8.1; Table 8.1).

Today, once-elastic U.S. industrial cities are encircled by a plethora of incorporated suburban municipalities. Meanwhile, elastic cities in the southeast and southwest are still able to expand their legal boundaries outward into unincorporated territory, thereby increasing in area and population. For instance, Phoenix's land area grew by 500 square miles from 1950 to 2010, from 17 to 517 square miles. Over this same time period, Atlanta grew from 37 to 133 square miles; Charlotte from 30 to 298; San Diego from 99 to 325; and Austin from 32 to 299. Table 8.2 compares these elastic and growing cities with several inelastic and industrial cities in the Northeastern and Midwestern United States.

Summary of the governmental framework in U.S. metropolitan regions

The U.S. federalist system guarantees that all geographic territories in the United States fall under the control of two layers of government: federal and state. The state level further divides its power between itself and a host of general and special purpose local governments. Because it is the state government that creates local governments, the latter are wholly subordinate to the state. In this vein, the state explicitly endows local governments with certain powers and no others, allows local governments to act relatively autonomously on local matters, or promotes some combination of these systems (see Box 8.2).

Regardless of the precise suite of powers granted to local governments by their parent states, one ability that tends to apply to all general purpose municipal governments is annexation. As long as the terms are mutually agreeable,

Table 8.2 Comparison of physical and population growth in some elastic and inelastic cities, 1950–2010

Inelastic Cities						
City	*1950*		*2010*		*Change*	
	Land Area	*Population*	*Land Area*	*Population*	*Land Area Change*	*Population Change*
Baltimore	78.7	949,708	80.9	620,961	2.2	–34.6%
Boston	47.8	801,444	48.3	617,594	0.5	–22.9%
Buffalo	39.4	580,132	40.4	261,310	1.0	–55.0%
Cleveland	75.0	914,808	77.7	396,815	2.7	–56.6%
Detroit	139.6	1,849,568	138.8	713,777	–0.9	–61.4%
Newark	23.6	428,776	24.2	277,140	0.6	–35.4%
Philadelphia	127.2	2,071,605	134.1	1,831,710	6.9	–11.6%
Pittsburgh	54.2	676,806	55.4	305,704	1.2	–54.8%
St. Louis	61.0	856,796	61.9	319,294	0.9	–62.7%
Youngstown	32.8	168,330	34.0	66,982	1.2	–60.2%
Elastic Cities						
Atlanta	36.9	331,314	133.2	420,003	96.3	26.8%
Charlotte	30.0	134,042	297.7	731,424	267.7	445.7%
Denver	66.8	415,786	153.0	600,158	86.2	44.3%
Houston	160.0	596,163	599.6	2,099,451	439.6	252.2%
Los Angeles	450.9	1,970,358	468.7	3,792,621	17.77	92.5%
Orlando	14.1	52,367	102.4	238,300	102.4	355.1%
Phoenix	17.1	106,818	516.7	1,445,632	499.6	1253.4%
Portland	64.1	373,628	133.4	583,776	69.3	56.2%
San Diego	99.4	334,387	325.2	1,307,402	225.8	291.0%
Tampa	19.0	124,681	113.4	335,709	94.4	169.3%

municipalities may annex unincorporated territories adjacent to them. This power led to tensions in the early twentieth century, whereby unincorporated suburban communities raced to form general purpose governments to avoid being annexed by central cities in the future. Eventually, this process caused most of America's older industrial cities to become inelastic and unable to expand their physical borders. At the same time, it contributed to a fragmented metropolitan government landscape, in which a large number of formal institutional actors occupy the same spatial extents.

In general, the more parties there are to a collective action problem, the higher the transaction costs, and the less (regional) cooperative alliances will be established. The high volume of local governments in U.S. metropolitan regions (see Fig. 8.1), therefore, is often derided as antithetical to the production of coordinated regional responses to the numerous urban problems facing shrinking central cities and their neighboring municipalities (Rusk 2013; OECD 2015). Accordingly, one strand of the new regionalism movement calls for supplanting decentralized local governments with a centralized regional government (Rusk 2013). To fully understand the implications of this proposal, it is first necessary to dig deeper into metropolitan structures of *governance*.

Metropolitan governance structures

Administratively fragmented metropolitan areas tend to generate elaborate governance structures that vary widely from place to place and across time (Oakerson 2004: 28). Thus, it is impossible to take and report on a census of governance frameworks in U.S. urban regions. In place of this momentous task, the current subsection begins by revisiting the debate from the new regionalism movement introduced earlier, over whether a centralized or decentralized governance solution is more amenable to intra-regional cooperation. Specifically, the centralization/decentralization debate is reframed as a preference over *monocentric* (single decision-maker) and *polycentric* (multiple decision-makers) models of governance (Ch. 7). However, this discussion reveals that the number of decision makers (one versus many) is only one dimension of regional governance solutions. Equally important is the degree to which regional decision-makers come together by way of self-organization or coercion. When this latter dimension is taken into account, it becomes clear that no monocentric-polycentric dichotomy exists in real-world regional governance – the toolbox of solutions is considerably fuller.

Monocentric and polycentric regional governance

Recall the two key ingredients of [uncoerced] inter-municipal collective action: (1) sufficiently high collective benefits and (2) low transaction costs, where *transaction costs* are impediments to establishing mutually beneficial alliances. These impediments come in many forms (Feiock 2007).

Two prominent forms of transaction costs found in metropolitan areas with shrinking cities are *information/coordination costs* and *division/negotiation costs*. To the extent that uncoerced regional cooperation requires high collective benefits, lack of information about the nature or possibility of benefits is a strong barrier to institutional collective action. Moreover, even where interdependence of institutions and the existence of mutual benefits are recognized, institutional leaders might disagree fundamentally on the level of effort and share of cooperative gains that their respective organizations are responsible for or entitled to, respectively. These *division/negotiation* costs also stand firmly in the way of regional cooperation. To illustrate how these abstract concepts become animated in real-world policy processes, Box 8.3 summarizes the recent case of water management in the shrinking city of Detroit, Michigan, and its surrounding region.

Box 8.3 Regional water management in Detroit, Michigan

Shortly after filing for bankruptcy in 2013, the city of Detroit began aggressive attempts to collect on outstanding debts owed to its municipal water department, the Detroit Water and Sewage Department (DWSD). At the time, the DWSD controlled an infrastructure network of nearly 1,100 square miles, including approximately 4,000 miles of water mains, and serviced the city of Detroit and neighboring suburban communities in eight counties. This expansive reach meant that the DWSD provided water to approximately 40 percent of the population of the state of Michigan (McCulloch 2013).

In response to the severe financial shortfalls Detroit was (and still is) experiencing, a state-appointed "emergency manager", Kevyn Orr, put forward a proposal to regionalize the DWSD. Orr's plan called for establishing a *regional water authority* (see the discussion of *special purpose local governments* in the preceding section) to provide water and sewer services to Detroit, its home county of Wayne, and the two neighboring counties of Oakland and Macombe. This so-called "regional plan" retained DWSD ownership of the infrastructure system that it was already managing. Hence, the proposed regional authority would be required to lease the system from DWSD under an arrangement that would raise much needed revenues for the city of Detroit – to the tune of $47 million per year for forty years.

Clearly, from Detroit's perspective, the possibility of long-term lease revenue was an attractive benefit. Importantly, suburban communities also stood to gain from a reconfiguration of the water provision system. As McCulloch (2013: 194) describes:

> [t]he relationship between Detroit and the suburban communities it serves has been contentious for decades, with suburban representatives questioning how water and sewer rates are developed [by DWSD] and how the funds generated are subsequently spent [by the city of Detroit].

In other words, the suburban municipalities that received water and sewer services from DWSD felt underrepresented in the existing city-controlled water management structure. In this way, the situation seemed pre-programed for a cooperative regional solution.

That being said, Orr's "regional proposal" met with notable backlash from suburban municipalities. The situation was summarized by *The New York Times* (Walsh 2014) as follows:

> people outside the city limits are not eager to help foot whatever bills may remain through new regional projects, taxes and fees. Prosperous, pro-business suburbs . . . have already spent years fighting plans to reinvent Detroit's convention center, art museum, zoo and bus system as regional enterprises.

The County Executive for neighboring Macombe County went so far as to tell National Public Radio (NPR) that, while he recognizes that the suburban communities in the county "need a healthy Detroit", there are limits on what he is willing to do to make a healthy Detroit happen. With respect to the regional water proposal, the county executive observed that suburban ratepayers saw the leasing arrangement as a "bad deal" that would not benefit communities outside of the city (http://www.npr.org/2014/03/09/287877060/city-versus-suburb-a-longstanding-divide-in-detroit).

Connecting theory with practice. In this example, the implementation of a regional solution faced at least two types of *transaction costs*. First, *information/coordination costs* created some distrust between suburban municipalities and the DWSD. Historically, suburban municipalities were kept out of DWSD decision-making processes. Such communities therefore lacked critical information about how rates were set and how revenues were spent (refer to the quote from McCullough above). Second, *division/negotiation costs* built a wedge between some suburban officials (refer to the preceding paragraph) and the backers of a regional plan in the city of Detroit. Namely, some suburban officials saw the proposal as a "bad deal" that would not give a fair share of benefits to ratepayers outside of the city.

Epilogue. In June 2015, the newly created Great Lakes Water Authority (GLWA) officially received a forty-year lease approval to control the DWSD infrastructure system. This new regional authority provides suburban municipalities with a stronger voice in water operations by giving them representation on the authority's governing board. The agreement will also raise $50 million per year in revenue for Detroit. The deal was opposed by the representative from Macombe County but supported by the remaining five members of the authority's board (Wisely and Guillen 2015). Days after the lease announcement, the *legitimacy* of the GLWA was challenged by a state legislator on the grounds that the agency was created "behind closed doors" during Detroit's federal bankruptcy hearings. This challenge, which is ongoing as of this writing, may prove problematic for the aforementioned lease agreement. As this chapter has detailed, local governments (including special purpose *authorities*) are creations of the state. Because the GLWA was not established by formal state-enabling processes, the state legislator's challenge might work to nullify the existing agreement (Cwiek 2015).

The Detroit water management case from Box 8.3 speaks to a common issue facing U.S. shrinking cities: the dissimilar positions and interests of relatively wealthy, better-off suburbs and resource-deficient central cities undermine the possibility of widespread inter-municipal cooperation. In the Detroit case, the creation of a regional special purpose local government (*viz.*, an authority) is being celebrated by some observers as a mechanism for aligning the interests of the suburbs and the city, thereby producing a mutually beneficial cooperative outcome. In a much broader context, this sort of approach is akin to what some participants in the new regionalism movement would like to see happen with *general purpose* government in metropolitan regions (Rusk 2013).

More precisely, the notion of consolidating existing interdependent communities into a centralized regional government is a form of **monocentric governance**. The monocentric governance model posits that the optimal means for achieving cooperation in the face of competing interests (e.g., city-suburb tensions) is to unite these interests under a common institutional authority (such as the Great Lakes Water Authority). Consolidation, it is argued, forces seemingly autonomous institutions to recognize their interdependencies, thereby "internalizing" any effects that their individual decisions might have on one another. The expected results of such arrangements are inter-municipal cooperation, and, by extension, greater efficiency in public service provision within the region (see Oakerson's 2004 discussion).

The counterpoint of monocentric governance is **polycentric governance** – a "process of decision-making whereby multiple independent actors interact to produce an outcome that is commonly valued" (Oakerson 2004: 21). Put differently, polycentricity is a *decentralized* mode of governance. It describes patterns of decisions that emerge from bottom-up interactions between a variety of institutions, both formal and informal. Recall from Chapter 6 that the "devolution" pillar of the American model of urban development was linked to intense inter-municipal competition for various forms of capital. While competitive outcomes definitely occur in U.S. metropolitan regions (Kantor 2010), competition is not the only process of urban development. Rather, institutions regularly cooperate in the absence of a centralized authority (Feiock 2004). In the shrinking city of Buffalo, New York, for example, "multiple government agencies and civic partnerships influence decision-making" (Cowell, Gainsborough, and Lowe 2015: 11). As a case in point, the efforts and activities of a local grassroots organization, People United for Sustainable Housing (PUSH Buffalo), led to the creation of a "Green Development Zone" (GDZ) on the west side of Buffalo (Cowell, Gainsborough, and Lowe 2015). The GDZ is a 25-square-block area that:

> combines green affordable housing construction, community-based renewable energy projects, housing weatherization projects, and green jobs training programs. Essentially, PUSH buys dilapidated properties, ensures that local laborers are hired to complete the project and receive job training, uses green technologies and weatherization to keep heating costs down for residents, and provides affordable rents to one of the most impoverished neighborhoods in the nation with a per capita income of approximately $9,000.
>
> (Owley and Lewis 2014: 256–257)

The GDZ was built from the ground up. Consistent with the established "conflict" model of community development (see Green and Haines 2015), PUSH Buffalo led a campaign against a New York State housing agency that was exercising control of vacant lots in western Buffalo for what were eventually found to be speculative purposes. The campaign resulted in a major public policy decision by New York State to invest millions of dollars into local neighborhood revitalization programs. This state-level policy eventually opened the door for PUSH to receive new financial resources and leverage its growing reputation to secure additional funding through philanthropic and related means. With this influx of financial capital, and following a series of participatory meetings and communicative planning processes, PUSH developed the GDZ. In its first three years of existence, the GDZ generated more than $6 million in investment, and its high profile successes led the city of Buffalo to invest another $350,000 in a neglected public space in the GDZ community (Dreier 2012).

The case of the PUSH Buffalo GDZ (see http://greendevelopmentzone.org/ for more information) illustrates quite well the distinction between *government* and *governance* articulated earlier in the chapter. Namely, one action that led causally to the establishment of the GDZ was PUSH's campaign against a state institution of *government*. This action made it such that an informal, nonprofit institution meaningfully perturbed the decision-making processes of a formal organization of government. Such outcomes are the essence of *governance* – the process by which human groups make trade-offs to protect the public realm (Oakerson 2004). Moreover, the case demonstrates that cooperative *governance* can occur without a coercive act of *government*. In the GDZ case, no state or local government set PUSH's policy agenda and priorities (i.e., green and affordable housing, quality local jobs, job training, revalorizing vacant land, and creating community wealth). Rather, the organization crafted its own mission and vision for Buffalo's west side, and its transformational impacts on the landscape have since led government organizations and policymakers to formally endorse and reinforce that vision (Dreier 2012; Cowell, Gainsborough, and Lowe 2015).

Although the preceding example focused on *vertical* cooperation at the scale of a shrinking city neighborhood – between informal community institutions of governance and formal local and state institutions of government – the lessons can usefully be applied to *horizontal* cooperation at the scale of a shrinking city's region. Formal and informal institutions interact in decentralized, intra-regional networks of governance on regular bases (see Box 8.3). Although these interactions sometimes generate deleterious competition between institutions at the same/horizontal level (e.g., Kantor 2010), when potential benefits are high and transaction costs are surmountable, decentralized networks also produce self-organized cooperative alliances that are capable of institutional collective action (Feiock 2013). Thus, the new regionalism debate over the degree to which regional approaches ought to be grounded in centralized or decentralized *government* is closely related to the debate over the relative efficacy of monocentric versus polycentric governance *structures* (Oakerson 2004). As the PUSH Buffalo case demonstrates, cooperative polycentric governance can emerge in the absence of a consolidated monocentric government (Feiock 2004). Nonetheless, both options may lead to intra-regional cooperation, depending on the nature and scope of the institutional collective action problem(s) and a host of situational and contextual factors (see Post 2004 for a discussion of these factors). At the same time, the implied dichotomy between centralized decision-making in a single regional government and decentralized decision-making in a polycentric governance network is a false one. In reality, a wider range of possibilities exists.

Alternative mechanisms of regional cooperation

In his influential work on the theory and policy implications of the institutional collective action problem, Richard Feiock (2009, 2013) argues that there are at least six general mechanisms for regional cooperation. Three of these mechanisms deal with bilateral arrangements, such as contracts between local governments and consultants, which are beyond the bounds of the current discussion. The remaining three mechanisms concern multilateral collective choice arrangements and are briefly outlined in this subsection.

The first general mechanism for facilitating intra-regional cooperation has already been mentioned: *consolidated* (i.e., monocentric) *regional governments/authorities*. Consolidated regional governments might take the form of *special purpose authorities*, or, less frequently, *general purpose governments*. With respect to the former, a specialized service area for which regional authorities are becoming increasingly popular is public transportation. For example, for many decades, the shrinking city of Cincinnati, Ohio, (Table 2.3) operated its own internal public transportation agency called the Cincinnati Transit Commission (CTC). In the mid-1970s, the CTC was dissolved and replaced by the Southwest Ohio Regional Transit Authority (SORTA), a state-created regional/special purpose government. SORTA is an *organized entity* that possess *governmental character* and *substantial autonomy* (refer to the discussion above). It is governed by a thirteen-member citizen board with seven members from the city of Cincinnati, three members from the city's parent county, and three members (one each) from the three adjacent counties. All board members are appointed by city and county public officials, and the authority is funded through local taxes, user fees, and federal sources (http://www.go-metro.com/about-metro/sorta). SORTA currently serves more than 280 communities in the Greater Cincinnati region (http://www.go-metro.com/riding-metro/communities-served).

Consolidated regional governments also sometimes take the form of general purpose local governments. Consider that from 1970 to 2000, the population in Louisville, Kentucky, decreased by more than 34 percent, from a peak of 390,639 down to 256,231. Like many other shrinking cities discussed in this book, by 2000, Louisville's metro region was characterized by a fragmented, and sometimes contentious, landscape of local government. Consistent with the American model of urban development, the eighty-some odd autonomous municipalities in Louisville's parent county of Jefferson were known for engaging in competition that led citizens, business leaders, and local developers to audibly call for an end to the pervasive "duplication and rivalry" (Greenblatt 2002). The region's solution, which had been proposed and rejected several times prior to its passage in a 2000 popular election, was a multilevel consolidation plan that merged the general purpose governments in the city of Louisville and the county of Jefferson into one, centralized, *regional government*. The new regional government putatively brought about "a more unified approach to economic development and governance" that officials have since credited with "attracting new businesses . . . [especially] in healthcare and logistics" (Wächter 2013: 19). In addition, Fitch Ratings, one of the three nationally recognized credit rating bureaus in the United States, cited the "diversified economy . . . of the newly formed metro government" in its decision to upgrade Louisville-Jefferson County to its highest AAA credit rating category in 2010 (Kelly and Adhikari 2013).

While the apparent *economic* gains from the Louisville-Jefferson County consolidation paint a positive picture of monocentric governance, Clarke (2006) cautions that centralization of decision-making authority in a single metropolitan government is likely to have unintended *social* consequences. Specifically, she found that the voting strength of African Americans in urban Louisville became significantly diluted in the comparatively suburban regional government. The propensity for consolidation to privilege the policy priorities and values of majority groups in this manner is a serious issue that cannot be overlooked in calls for local government consolidation.

The second general regional governance mechanism considered here is characterized by a marginally more democratic and voluntary structure compared to a regional government. Explicitly, a *regional organization* is an association of either formal, informal, or both formal and informal institutions that generally does not possess *governmental character*. To the extent that local governments participate in regional organizations, they tend to retain some or all of their autonomy (Feiock 2009). In this way, unlike regional governments, regional organizations typically lack the power to coerce members to do things they do not want to do. Further, regional organizations cannot coerce nonmembers to become members. Even so, the voluntary membership bases of regional organizations quite regularly bind themselves to agreements that they uphold to avoid intra-organizational sanctions. Whereas these sanctions can be reputational or otherwise unexacting from a financial perspective, occasionally they are linked to the loss of funding. That is, regional organizations are sometimes formed by way of state- or federal-level financial incentives to promote regional cooperation. Where municipal governments enter into state-sponsored regional organizations, and they subsequently fail to follow the rules of participation in those organizations (both self-imposed rules and those established by the incentivizing authority), they may be ineligible to receive anticipated funds or future funds in the given program area.

An example of a state-sponsored regional organization is the Mohawk Valley Sustainability planning project (MVSPP) in upstate New York (http://www.sustainablemohawkvalley.com/). Backed by portions of a "$100 million competitive grant program to encourage communities to develop regional sustainable growth strategies", the MVSPP is an association of representatives from six counties, officials from a regional economic development authority, and technical assistants from a local environmental planning firm. The remits of the organization, whose members voluntarily contracted with one another to establish the MVSPP, are to:

- establish a statewide sustainability planning framework that will aid in statewide infrastructure investment decision making;
- outline specific and tangible actions to reduce greenhouse gas emissions consistent with a goal of 80% carbon reductions by the year 2050;
- inform municipal land use policies;
- serve as a basis for local government infrastructure decision making;
- help guide infrastructure investment of both public and private resources; and
- provide each region with a sustainability plan that will enable them to strategically identify and prioritize the projects they submit for consideration to the Implementation Grant stage (http://www.sustainablemohawkvalley.com/).

Regional organizations such as the MSVPP, despite not having the corporate powers of formal institutions of government, are often very successful at marshalling resources in pursuit of institutional collective action. However, when funding for such organizations or participation therein is tied to a mandate rather than an incentive, existing forms of intra-regional cooperation are occasionally "crowded out" (Kwon and Park 2014). Stated another way, the creation of the second general type of regional institution considered in this section might bring about the destruction of the third type: *collaborative groups and regional councils*.

Collaborative groups and regional councils are informal networks in which shared norms of trust and reciprocity (see the discussion of "social capital" in Chapter 4) facilitate the spread of information, and otherwise work to reduce the transaction costs that might stand in the way of inter-municipal collective action (Feiock 2009). Collective decisions and other agreements reached in such networks rely on mutual consent, as collaborative groups tend to be self-organized institutions of governance with neither formal authority nor formal (e.g., state financed) sponsorship. Nevertheless, these voluntary associations, insofar as they produce shared understandings and expectations about regional challenges and opportunities, are highly efficacious means for achieving regional cooperation.

As an example, in the metropolitan region that includes the once-shrinking city of Milwaukee, Wisconsin (Beauregard 2009), local development officials informally agreed to share information with each other on conversations they have with any regional businesses that are seeking location incentives to move their current operations (Feiock 2009: 365). Within the conventional, competitive American model of urban development, such an agreement is irrational. Namely, if an official in city A is approached by a business currently located in city B about relocation, then the official in A should attempt to poach the business and grow its own economy at the expense of city B. Contrary to this logic, members of the informal group of development officials in the Milwaukee metro actively coordinate to avoid these win/lose outcomes and thus to promote the health of the region over their individual self-interests.

Figure 8.2 places the three general regional cooperation mechanisms discussed in this subsection onto a two-dimensional graph. The vertical axis of the graph captures the extent to which cooperative arrangements are binding (i.e., members cannot withdraw from them without cost) or nonbinding (i.e., members can exit at little or no cost). The horizontal axis measures the degree to which institutions are formed by coercion, incentivization, or self-organization. In general, *regional authorities* and consolidated governments tend to create binding agreements through coercive mechanisms. In the case of a regional water authority, for example (see Box 8.3), the authority is established by state or federal enabling legislation, and its members enter into enforceable contracts. Consolidated regional governments are created either through state enabling legislation or, as was the case in Louisville, a popular referendum election in which voters exercise their decision-making authority. *Regional organizations* are often, but not always, comprised of voluntary members. However, the members frequently establish rules of conduct and contracts that are internally enforceable. When such organizations are formed as a result of state- or federal-level incentives for regional cooperation, as in the case of the Mohawk Valley Sustainability planning project, collective agreements may also be enforceable at a higher (state) level of government. Finally, *collaborative groups* tend to be self-organized, voluntary, informal associations that enter into collective arrangements and agreements, which are enforced exclusively through customs and norms.

Note that Figure 8.2 is intended as a visual aid for thinking about the various mechanisms of regional cooperation. Real-world examples of these types of institutions may fall in different locations, or may span more territory, in the graph. The placements used in the figure should therefore be considered relative, not absolute. In general, one may think of institutions nearer to the origin of the graph

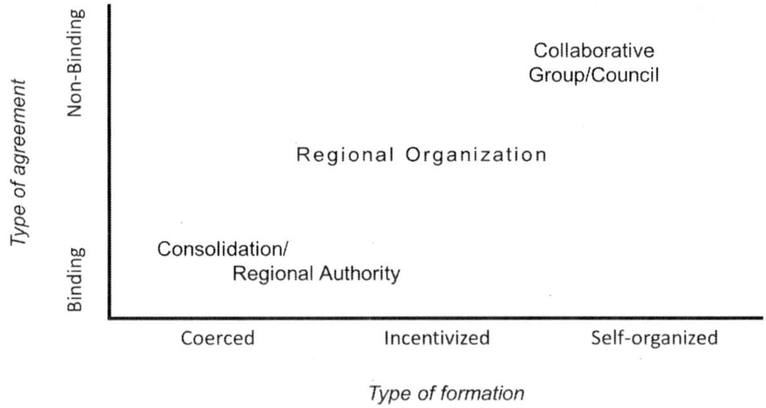

Figure 8.2 Mechanisms of regional cooperation (inspired by Feiock 2009: 360)

as being closer to *monocentric governance* structures, and institutions more distant from the origin in both dimensions as relatively more representative of *polycentric governance structures*. Once again, however, this statement is a generality that should not be interpreted as a global property of regional governance institutions.

Cooperation in regional land use

To end this chapter, it is useful to apply the above lessons to a particularly thorny issue in metropolitan regions with shrinking cities: coordinated regional land use. Local governments are the primary public superintendents of private land and real property use in the United States. That is, state governments authorize their local government subsidiaries to regulate and manage land use and building practices within the latter's jurisdictional boundaries (Platt 2014). The broadest and arguably most popular technique for controlling land use is zoning. Classic land use zoning, also known as **Euclidean zoning** in recognition of the court case that held the practice to be constitutional (*Village of Euclid v. Ambler Realty Co.*), is a comprehensive verbal (zoning *code*) and cartographic (zoning *map*) specification of permitted and prohibited uses for all parcels of land within the borders of a given municipality.

Euclidean zoning is part of a general purpose local government's *regulatory* or *police power*. That is, zoning codes and zoning maps are intended to protect the public health, safety, and welfare, presumably by minimizing harmful social and social-environment land use interactions within a jurisdiction. With that objective in mind, zoning codes are ideally preceded by, and justified within, comprehensive plans. A **comprehensive plan** is a document and supporting materials written to guide future municipal development. Such plans are forward-looking, and provide visions for municipalities within general frameworks that are situated in the appropriate local, state, and federal policy contexts. That said, in practice, zoning tends to occur either before or entirely divorced from comprehensive planning (Platt 2014). Among other reasons, this absence of planning has led critics of Euclidean zoning to argue that it functions more as a mechanism for protecting narrow, parochial interests than the public interest at large. Specifically, critics point to at least three zoning practices that contribute to the "balkanization" of U.S. metropolitan regions, and which tend to have polarizing social effects (Platt 2014: 200–201):

1 *Exclusionary zoning* has been employed by relatively wealthy (especially suburban) municipalities to require minimum residential lot sizes, limit the availability of smaller homes, and prevent apartment or public housing development. In all of these cases, the result is often that households with modest financial means are excluded from moving into such municipalities.
2 *Fiscal zoning*, which also tends to be linked to wealthier suburban communities, involves zoning regulations that minimize local property taxes by simultaneously incentivizing activities like shopping centers and industrial parks that generate municipal revenue, while disincentivizing activities like affordable housing that subtract from municipal revenue.

3 *NIMBYism ("not in my backyard")* occurs when municipal institutions of *governance* use zoning regulations and other legal means to block locally unwanted land uses (e.g., mental health facilities) from occurring in their communities.

The social polarization implicated by these three zoning practices regularly manifests as spatial polarization, with newer suburbs effectively "zoning out" undesirable land uses attributed to older cities and first-ring suburbs (e.g., low-quality housing, industrial buildings, etc.). Thus, instead of working with struggling municipalities to improve conditions in the overall region, zoning and other legal mechanisms enable thriving municipalities to simply [attempt to] close off their territories to undesirable "urban" conditions. Using the terms defined earlier in this chapter, Euclidean zoning therefore contributes to the institutional collective action problem faced by municipalities in regions with shrinking cities.

Crucially, land use policies such as zoning regulations tend to be rigid and persistent over time, even when they are not functioning as originally intended (Weaver and Knight 2014). Completely replacing existing zoning codes with new "green zoning" institutions such as those discussed in Chapter 7 is therefore a lengthy and uphill political process in one jurisdiction, let alone in all the jurisdictions of a metropolitan region. As a consequence, alternative strategies for coordinating land and real property uses across jurisdictions are needed. One strategy that is receiving more and more attention in the shrinking cities community is a **land bank**. Land banks are:

> public or nonprofit entities created by local governments to acquire, manage, maintain and repurpose vacant and foreclosed buildings and lots. [They] are not financial institutions; they are single-purpose entities charged with finding new and meaningful uses for abandoned properties.
>
> (Kildee 2010: 4)

While land banks vary in their powers and levels of activity from state to state and region to region, they are generally aimed at solving problems related to vacant and abandoned properties. Moreover, to the extent that land banks require state enabling legislation (LaCroix 2011) and subsequent authorization by local governments (see the passage quoted above), they are hallmarks of inter-governmental cooperation. Almost invariably, land banks take the form of regional *authorities* (Fig. 8.2). Unlike regional authorities that are established by state mandate or coercion, however, most U.S. land banks tend to be *self-organized*: local governments come together to request or propose that one be created (Hackworth 2014).

The model Genesee County Land Bank (GCLB) in Flint, Michigan, is an extreme case of this self-organized institutional cooperation. Namely, the GCLB was founded *two years before* the state of Michigan officially passed legislation authorizing the creation of land banks. Prior to this legislation, the GCLB organized in a comparatively informal "urban cooperative agreement" aimed at assembling abandoned properties for productive reuse (Alexander 2008: 23). Following passage of the state's land bank legislation, the GCLB was then formalized as an institution of *regional government*. Recognizing that the problems of "urban" shrinkage and decline extend beyond central city borders, the member governments of the GCLB now work closely to create new, affordable housing opportunities in the region, demolish and rehabilitate problem properties, limit property speculation, and prevent foreclosure where possible (Hackworth 2014). Since its inception, the GCLB has contributed to property value increases in excess of $100 million and has catalyzed reuse of over 1,000 brownfield properties (Alexander 2008).

The broad techniques that land banks use toward these and related ends include: (1) taking title to tax delinquent properties; (2) working with municipal, county, and sometimes state agencies to "clean titles" by erasing liens; (3) planning for the long-term use of tax delinquent properties (i.e., coordinating the reuse of such properties toward a regional goal); (4) maintaining properties; and, ultimately, (5) transferring properties to approved owners who will use the properties in ways that comport with planned regional goals (LaCroix 2011).

Concluding remarks

In framing the barriers to institutional collective action and defining the various organizational entities that participate in metropolitan government and governance, this chapter kept to a single theme: whether state-mandated or self-organized, inter-governmental cooperation is a necessary condition for effectively addressing the challenges of shrinking cities. Importantly, inter-municipal cooperation can be achieved through a variety of mechanisms and does not always hinge on the consolidation of local governments into a single regional government (contra Rusk 2013). Nor, however, do decentralized governance systems always produce cooperation. Researchers, practitioners, and government officials need to work together to recognize the most socially beneficial regional governance for the specific context under consideration.

Regardless of contextual circumstances, a general heuristic is that the most successful regional cooperative arrangements exist where potential benefits from collective action are high, and transaction costs are low. Enhancing efforts to educate the parties to an institutional collective action problem about the nature of mutual benefits from cooperation therefore seems to be a valuable near-term priority for, among others, urban planners in shrinking cities who wish to intervene in restructuring the governance in existing regional planning systems. Similarly, facilitating communication and otherwise bridging connections between parties might have utility for reducing transaction costs that stand in the way of cooperation.

One topical domain in which planners are increasingly emphasizing broad-scale benefits and building up lines of communication between metropolitan (and global) stakeholders is that of sustainability. Specifically, the notions of sustainability and resilience are rapidly ascending the priority lists of government and governance organizations in urban America in ways that emphasize the interdependence of places and peoples (and, hence, the value of collective action). The next chapter explores how these issues are becoming animated in the shrinking cities discourse and incorporated into new ways of thinking that challenge prevailing pro-growth oriented approaches to solving urban problems.

Note

1 See: http://www.census.gov/govs/

References

Alexander, Frank S. 2008. "Land Banking as Metropolitan Policy." Metropolitan Policy Program, Washington, DC, Brookings Institution. http://www.smartgrowthamerica.org/documents/brookings_land_banking.pdf

Badger, Emily. 2015. "What Happens When a Metropolitan Area Has Way Too Many Governments?" *The Washington Post*, February 18, 2015, Wonkblog, https://www.washingtonpost.com/news/wonk/wp/2015/02/18/what-happens-when-a-metropolitan-area-has-way-too-many-governments/

Beauregard, Robert A. 2009. "Urban Population Loss in Historical Perspective: United States, 1820–2000." *Environment and Planning A* 41 (3):514–528.

Bluestein, Frayda S. 2006. "Do North Carolina Local Governments Need Home." *North Carolina Law Review* 84 (6):1983–2030.

Clarke, Kristen. 2006. "Voting Rights & City-County Consolidations." *Houston Law Review* 43(6): 621–699.

Cowell, Margaret, Juliet F. Gainsborough, and Kate Lowe. 2015. "Resilience and Mimetic Behavior: Economic Visions in the Great Recession." *Journal of Urban Affairs* 38 (1):61–78.

Cwiek, Sarah. 2015. "One State Rep Says New Detroit Regional Water Authority Needs Legislative Approval." Michigan Radio. http://michiganradio.org/post/one-state-rep-says-new-detroit-regional-water-authority-needs-legislative-approval#stream/0

Dewar, Margaret, and June Manning Thomas, eds. 2012. *The City after Abandonment*. Philadelphia: University of Pennsylvania Press.

Dreier, Peter. 2012. "Rust Belt Radicals: Community Organizing in Buffalo." *New Labor Forum* 21 (2):100–104.

Favro, Tony. 2010. "American Cities Seek to Discover Their Right Size." *City Mayors*. http://www.citymayors.com/development/us-rightsizing-cities.html

Feiock, Richard C. 2004. "Politics, Institutions and Local Land-Use Regulation." *Urban Studies* 41 (2):363–375.

Feiock, Richard C. 2007. "Rational Choice and Regional Governance." *Journal of Urban Affairs* 29 (1):47–63.

Feiock, Richard C. 2009. "Metropolitan Governance and Institutional Collective Action." *Urban Affairs Review* 44 (3):356–377.

Feiock, Richard C. 2013. "The Institutional Collective Action Framework." *Policy Studies Journal* 41 (3):397–425.

Feiock, Richard C., Jill Tao, and Linda Johnson. 2004. "Institutional Collective Action: Social Capital and the Formation of Regional Partnerships." In *Metropolitan Governance: Conflict, Competition, and Cooperation*, edited by Richard C Feiock, 147–158. Washington, DC: Georgetown University Press

Frug, Gerald E. 1984. "The City as a Legal Concept." In *Cities of the Mind: Images and Themes of the City in the Social Sciences*, edited by Lloyd Rodwin and Robert M. Hollister, 233–290. Boston: Springer US.

Grassmueck, Georg, and Martin Shields. 2010. "Does Government Fragmentation Enhance or Hinder Metropolitan Economic Growth?" *Papers in Regional Science* 89 (3):641–657.

Green, Gary Paul, and Anna Haines. 2015. *Asset Building & Community Development*. Thousand Oaks, CA: Sage Publications.

Greenblatt, Alan 2002. "Anatomy of Merger Greater Louisville Is About to Be Born: How Much Greater Will It Be?" e.Republic. http://www.governing.com/topics/politics/Anatomy-Merger.html.

Hackworth, Jason. 2014. "The Limits to Market-Based Strategies for Addressing Land Abandonment in Shrinking American Cities." *Progress in Planning* 90:1–37.

Hardin, Garrett. 1968. "The Tragedy of the Commons." *Science* 162 (3859):1243–1248.

Kansas, Atkin V. 191 U.S. 207, 208 [1903].

Kantor, Paul. 2010. "City Futures: Politics, Economic Crisis, and the American Model of Urban Development." *Urban Research & Practice* 3 (1):11.

Kelly, Janet M., and Sarin Adhikari. 2013. "Indicators of Financial Condition in Pre- and Post-Merger Louisville." *Journal of Urban Affairs* 35 (5):553–567.

Kildee, Dan. 2010. "Land Banks as a Redevelopment Tool." Federal Reserve Bank of Philadelphia. https://www.philadelphiafed.org/community-development/publications/cascade/75/04_land-banks-as-redevelopment-tool.

Kübler, Daniel, Urs Scheuss, and Philippe Rochat. 2012. "Place Equality Regimes in Swiss Metropolitan Areas." The 22nd World Congress of the International Political Science Association, Madrid, July 8–12.

Kwon, Sung-Wook, and Sang-Chul Park. 2014. "Metropolitan Governance: How Regional Organizations Influence Interlocal Land Use Coordination." *Journal of Urban Affairs* 36 (5):925–940.

LaCroix, Catherine J. 2011. "Urban Green Uses: The New Renewal." *Planning & Environmental Law* 63 (5):3–13.

Lee, B., P. Gordon, J.E. Moore, and H.W. Richardson 2011. "The Attributes of Residence/Workplace Areas and Transit Commuting." *Journal of Transport and Land Use* 4 (3):43–63.

Leo, Christopher, and Kathryn Anderson. 2006. "Being Realistic about Urban Growth." *Journal of Urban Affairs* 28 (2):169–189.

Lowery, D. 2000. "A Transactions Costs Model of Metropolitan Governance: Allocation Versus Redistribution in Urban America." *Journal of Public Administration Research and Theory* 10 (1):49–78.

Mallach, Alan, and Levea Brachman. 2013. *Regenerating America's Legacy Cities*. Cambridge: Lincoln Institute of Land Policy.

McCulloch, J.P. 2013. "Detroit Water and Sewerage Department: Bringing Credibility to a Beleaguered System through Regional Cooperation and Technology." *Journal of Law in Society* 14:193–206.

Morgan, Bronwen, and Karen Yeung. 2007. *An Introduction to Law and Regulation: Text and Materials*. Cambridge, UK: Cambridge University Press.

Oakerson, Ronald J. 2004. "The Study of Metropolitan Governance." In *Metropolitan Governance: Conflict, Competition, and Cooperation*, edited by Richard C. Feiock, 17–45. Washington, DC: Georgetown University Press.

Orfield, Myron. 1997. *Metropolitics: A Regional Agenda for Community and Stability*. Washington, DC: Brookings Institution Press.

Organisation for Economic Co operation and Development. 2015. *The Metropolitan Century: Understanding Urbanisation and Its Consequences*. Paris: OECD Publishing.

Ostrom, Elinor. 2005. *Understanding Institutional Diversity*. Princeton: Princeton University Press.

Ostrom, Elinor. 2008. "Polycentric Systems as One Approach for Solving Collective-Action Problems." http://papers.ssrn.com/sol3/papers.cfm?abstract_id=1936061.

Owley, Jessica, and Tonya Lewis. 2014. "From Vacant Lots to Full Pantries: Urban Agriculture Programs and the American City." *University of Detroit Mercy Law Review* 91:233–258.

Platt, Rutherford H. 2014. *Land Use and Society: Geography, Law, and Public Policy*. 2nd ed. Washington, DC: Island Press/Center for Resource Economics.

Post, Stephanie S. 2004. "Metropolitan Area Governance and Institutional Collective Action." In *Metropolitan Governance: Conflict, Competition, and Cooperation*, edited by Richard C. Feiock, 67–92. Washington, DC: Georgetown University Press.

Richardson, Jesse J. 2011. "Dillon's Rule Is from Mars, Home Rule Is from Venus: Local Government Autonomy and the Rules of Statutory Construction." *Publius: The Journal of Federalism* 41 (4):662–685.

Richardson, Jesse J., Meghan Zimmerman Gough, and Robert Puentes. 2003. *Is Home Rule the Answer? Clarifying the Influence of Dillon's Rule on Growth Management*. Washington, DC: Brookings Institution Center on Urban & Metropolitan Policy.

Rusk, David. 2013. *Cities without Suburbs: A Census 2010 Perspective*. Washington, DC: Woodrow Wilson Center Press.

Rybczynski, Witold, and Peter D. Linneman. 1999. "How to Save Our Shrinking Cities." *Public Interest* 135:30–44.

Savitch, H.V., and Ronald K. Vogel. 2000. "Introduction: Paths to New Regionalism." *State & Local Government Review* 32 (3):158–168.

United States Census Bureau. 2013. "Individual State Descriptions: 2012." In *2012 Census of Governments*. Washington, DC: United States Census Bureau.

Village of Euclid v. Ambler Realty Co. 272 U.S. 365 [1926]. US Supreme Court.

Wächter, Petra. 2013. "The Impacts of Spatial Planning on Degrowth." *Sustainability* 5 (3):1067–1079.

Walsh, Mary Williams. 2014. "Detroit's Plan to Profit on Its Water, by Selling to Its Neighbors, Looks Half Empty." *The New York Times*, A11.

Weaver, Russell C., and Jason Knight. 2014. "Evolutionary Mismatch as a General Framework for Land Use Policy and Politics." *Land* 3 (2):504–523.

Wheeler, Stephen M. 2002. "The New Regionalism: Key Characteristics of an Emerging Movement." *Journal of the American Planning Association* 68 (3):267–278.

Wisely, John, and Joe Guillen. 2015. "Great Lakes Water Authority OKs Lease of Detroit System." *Detroit Free Press*, June 12, 2015.

9 Sustainability and resilience

In Chapter 7, we observed that alternative ways of approaching urban development from the *rightsizing* and *smart decline* planning movements are beginning to supplant prevailing *pro*-growth approaches in certain U.S. shrinking cities. Within these paths, growth is not tantamount to "winning" as it has historically been portrayed (see Ch. 6), and shrinkage should not be trivialized or avoided at all costs (see Hospers 2014). Rather, *rightsizing* frames population shrinkage as an opportunity to experiment with new urban fabrics that might make affected cities more "sustainable" over time (Schilling and Logan 2008). The rightsizing movement coincides with a modern surge in urban scholarly research that conceptualizes cities as complex, dynamic, and evolutionary systems (Marshall 2009; Batty 2013), and as such, the language of systems, complexity, and evolution are beginning to see use in the professional and scholarly discourses of urban planning and urban change (Eraydin and Taşan-Kok 2013; Pickett, Cadenasso, and McGrath 2013). Hence, terms like *sustainability* and *resilience* now feature in planning tools and strategies in shrinking cities, including those that emanate from prevailing pro-growth mental models. This chapter begins by presenting working definitions for *resilience*, *sustainability*, and other related terms. From there, we describe several examples of how urban sustainability and resilience have been pursued in shrinking cities. In doing so, we compare policy instruments to provide understandings of sustainability and resilience within the existing broader context of urban development.

Vulnerability, stability, resilience, and sustainability

From a shrinking cities perspective, the concepts of *vulnerability*, *stability*, *resilience*, and *sustainability* have considerable utility, in that they deal with the extent to which a system (such as a city or neighborhood) is susceptible, and possesses the capacity to respond, to changes that could alter its identity (de Vries 2013).

The notion that cities and their entities (actors, spaces, built structures, etc.) are susceptible to negative change (i.e., *decline*) is the domain of vulnerability. **Vulnerability** refers to "exposure to risks" and the "limited capacity of [systems or entities] to avoid or absorb the harm" brought about by those risks (Taşan-Kok, Stead, and Lu 2013: 71). Contained in this definition are three dimensions of vulnerability: exposure, sensitivity, and adaptive capacity (Adger and Brown 2009). Exposure concerns the type and magnitude of stress faced by a given [part of a] system. Sensitivity captures the extent to which a [part of a] system "can absorb the impacts [of stress] without suffering long-term harm or some significant state change" (Adger and Brown 2009: 110). Finally, adaptive capacity relates to the ability of a [part of a] system to evolve – via individual-level behavioral adaptations – to accommodate stress (Page 2011). Standing up to "stress" is clearly an important idea – and strategies to improve a city's potential in this respect include reducing its exposure to stress, decreasing its sensitivity to stress, strengthening its adaptive capacity (through agent-level behavioral modifications), or all of these. What matters, then, is whether a city has, or does not have, the ability to "survive" a given disturbance.

Three distinct concepts correlate with this ability to "survive". In the first place, **stability** refers to satisfying the three conditions of constancy, persistence, and reversibility (de Vries 2013). In other words, a stable system is one that remains essentially unchanged, persists through time, and returns to an equilibrium state following a disturbance (Page 2011). Stated somewhat more simplistically, *stability* refers to maintaining or restoring a previously realized state of a system, such as a thriving downtown core. The next section argues that decision-makers in shrinking cities have historically sought to make systems stable in this sense, often through attempts to (re)grow the sizes of their populations and economies via the types of traditional growth-oriented policy and planning instruments discussed in Chapter 6.

Second, a system is said to have **resilience** if it can absorb a disturbance without shifting to a different regime or qualitative state (Holling 1973). That is, *resilience* is the "capacity of a system to experience shocks while retaining essentially the same functions . . . and, therefore, identity" (Eraydin and Taşan-Kok 2013: 5). Although this description sounds close to the preceding definition of *stability*, there is a critical distinction here. Stability relates to a system's attraction to a particular (i.e., single) equilibrium *state*. In contrast, resilience concerns a system's performance of a particular *function* in the face of a certain event, where a single function can be produced in many states. That is, a resilient system need not return to its same pre-disturbance (or any) equilibrium (Page 2011). Rather, such a system copes with the circumstances – it adapts and self-organizes to maintain its functions in a changed, post-disturbance environment (Adger and Brown 2009). A resilient system therefore does not have to satisfy the constancy and reversibility conditions necessary for stability. These notions of resilience and stability are gaining increasing popularity in urban planning (Pickett, Cadenasso, and Grove 2004; Pickett, Cadenasso, and McGrath 2013).

The third concept that correlates with a system's ability to "survive" is the one that receives the most attention but is the most controversial to define: *sustainability*. Sustainability is a contested concept because it is framed and reframed by an endless constellation of actors who view it from different perspectives, often in ways that correspond to political ends linked to pro-growth mental models (Voinov 2008; Mansfield 2009). Engaging with these myriad issues goes beyond the scope of this chapter. Consequently, although social dimensions and applications of sustainability are considered below for their relevance to planning in shrinking cities, here we extract working definitions of sustainability strictly from the primary complex systems literature to maintain consistency. Along those lines, **sustainability** is "the capacity to create, test, and maintain adaptive capability" (Holling 2001: 390). Recall that a resilient system possesses adaptive capacity that insulates it from stress. A sustainable system is thus a resilient system that can produce *new* adaptive capacity, test that capacity against disturbances, and retain the qualities that make it successful in the face of new, more diverse disturbances. Put differently, the adaptive capacity of a sustainable system *evolves* to make the system more resilient over time. This observation, in turn, means that a sustainable system has the ability to function "long into the future" (Common and Stagl 2005: 8). It further implies that "resilience is the key to . . . sustainability". In order to be sustainable, a system must be resilient, but it need not be stable (Wu and Wu 2013: 219).

Applications to shrinking cities: sustainability and sustainable development

Cities are inherently *vulnerable* to disturbances – natural disasters, deindustrialization, suburbanization, decline, etc. – and, presumably, local decision-makers are responsible for guiding municipalities through (i.e., "surviving") these disturbances. Whereas the interest in protecting cities against vulnerabilities long predates the incorporation of complex systems vocabulary into urban studies lexicons (e.g., Platt 2014), the "new" language of sustainability and resilience provides modern participants in urban change discourses with a powerful set of metaphors (Pickett, Cadenasso, and Grove 2004). The first contemporary wave of these metaphors followed the 1987 United Nations World Commission on Environment and Development (WCED) report titled *Our Common Future*, also called the Brundtland Report (World Commission on Environment and Development 1987). Concerned with the perceived conflict between economic development and environmental protection, the Brundtland Report coined the term **sustainable development** to mean economic development that "meets the needs of the present generation without compromising the ability of future generations to meet their own needs" (World Commission on Environment and Development 1987: 43).

This definition of sustainable development, and the Brundtland Report more broadly, called attention to environmental issues and promoted intra- and intergenerational equity on the global [urban] political agenda. However, by doing so in the context of development, the quest for "sustainability" – at least in the United States – eventually became one for environmentally sensitive economic *growth*. The resultant metaphor of "sustainable cities" was thus constructed atop the prevailing pro-growth ideology (Bulkeley and Betsill 2005). Indeed, this "new policy agenda" of sustainability (Agyeman, Bullard, and Evans 2002) painted a lasting picture that "economic growth is good for [both] people and the environment" (Mansfield 2009: 38). More explicitly,

> [t]he dominant message is that sustainable development is a productive answer to the world's economic and environmental problems, it provides a way for opposing groups to come together, and it is . . . framed in such a way that economic growth is . . . the most desirable way to achieve human development.
> (Lewis 2000: 257)

While these arguments are drawn in rather broad strokes that fail to capture finer-grained detail, they nevertheless reflect a growing consensus that ecological notions of "sustainability" have been co-opted by adherents to the American model of urban development (Ch. 6) for the purpose of "greening" neoliberal policy instruments and "maintaining [focus on] economic growth" (Cook and Swyngedouw 2014: 173). Stated another way, sustainability has been popularly framed as a technical systemic attribute that is acquired not by sacrificing a commitment to growth, but by shifting to more ecologically responsible modes of production and consumption – for instance, substituting decreasing stocks of natural capital with synthetic alternatives and operating more fuel efficient vehicles (see Mansfield 2009).

Importantly, this perspective gives way to a "market-oriented" approach to sustainability in cities (Greenberg 2013), whereby familiar American model of urban development policy instruments (Ch. 6) undergo "green-washing" (Cook and Swyngedouw 2014). For instance, "signature" projects are built according to Leadership in Energy and Environmental Design (LEED) standards; vacant lots are temporarily maintained as "green spaces" that store value for future economic growth; or demolition programs recycle structural materials, among other things (Wheeler and Beatley 2014). In this manner, sustainable development keeps intact the same pro-growth *mental model* that was discussed in Chapter 6 – the "new policy agenda" is new in its name and packaging but not in its core attitudes, values, beliefs, norms, and conventions regarding economic development (Agyeman, Bullard, and Evans 2002; Zovanyi 2013).

Examples of these circumstances abound, though they are sometimes masked by their rhetoric and good intentions (Cook and Swyngedouw 2014). Here, we single out two recent projects, but the aim is not to cast them in a negative or unproductive light; the intent is only to illustrate how their engagements with "sustainability" are fundamentally situated in the growth-first American model of urban development. First, the city of Hartford, Connecticut, has contracted in population by approximately 30 percent from its peak value in 1950 to the most recent (2010) decennial census. As a former manufacturing city, the global and regional post-World War II economic shocks that were discussed in Chapter 5 created a citywide perception of shrinkage and decline in Hartford. Many of the city's downtown businesses and residents pulled out of the area as post-industrial economic conditions trended downward. And these patterns of social and economic capital out-migration led to central city vacancy and qualitative decline (Ch. 4).

Around 2008, a space known as Constitution Plaza East, which is located "at the front door of the city's downtown near the [waterfront]", was targeted for redevelopment (City of Hartford 2009: 1). Consistent with the American model, this project reasoned that a strategy of substantial and high-profile downtown reinvestment, including a "signature building" (City of Hartford 2008: 1), might kick-start a chain reaction of local economic development (Ch. 6). Unlike earlier variants of the large-scale urban development project, however, the redevelopment plan for Constitution Plaza East was built on a foundation of sustainability. Specifically, the guiding principles of the redevelopment plan were derived from the **Smart Growth** movement in urban planning (City of Hartford 2008: Preface), which was detailed in Chapter 7.

In fostering the preservation of "open space, natural areas, and farmlands", "maintaining historic investment", and developing "metropolitan areas with a decreasing emphasis on the automobile" (Warner 2006: 169), the Smart Growth movement seeks to weave together compact urban fabrics that discourage so-called *unsustainable* patterns of growth, namely sprawl. In this way, the Constitution Plaza East plan sought (seeks) to achieve sustainability via high-density, mixed-use "development with office space, residential units, and ground floor retail, all located within high-rise [signature] buildings" (City of Hartford 2008: 1). Of course, the ideology that underlies this strategy tends to assume that a positive supply-side shock (i.e., increasing the stock of downtown residential and commercial space) will bolster the demand for downtown real estate – which is a questionable assumption anywhere, let alone in weak market, shrinking cities (Schilling and Mallach 2012).

As its name implies, the Smart Growth movement is oriented toward economic and population (re)growth. Indeed, most Smart Growth legislation requires "upzoning to accommodate future [increased] populations" (Warner 2006: 175–176). For that reason, in contrast to exercises in *smart decline* (Ch. 7), Smart Growth and its roll out in "signature" projects and other American-style urban policies (Ch. 6) does not necessarily increase a city's capacity to "create, test, and maintain adaptive capability" (Holling 2001: 390). Hence, drawing on the terminology from above, in a shrinking cities context, Smart Growth and related "sustainable developments" may ultimately privilege *stability* (i.e., returning to pre-shock levels of population and economic activity) over *sustainability* (i.e., increasing resilience to future and more diverse ranges of shocks).

A second example of "pro-growth sustainable development" (Zovanyi 2013) can be seen (perhaps paradoxically) in the Cleveland EcoVillage. A poster-child of shrinking cities, the one-time manufacturing hub of Cleveland, Ohio, had a population in 2010 (396,815) that was less than half of its peak (1950) value of 914,808 people. The visually and psychologically prominent patterns of shrinkage in Cleveland have generated "many new urban sustainability ideas" and "shifts in thinking" (Wheeler and Beatley 2014: 424). While other authors have explored several of Cleveland's sustainability and revitalization policies in greater detail (LaCroix 2011), the focus here is narrowly on "perhaps the most tangible new [as of 2004] reflection" of the city's sustainability initiatives (Wheeler and Beatley 2014: 424). The Cleveland EcoVillage is a "model urban village that will realize the potential of urban life in the most ecological way possible" (Scott 2002). Based on Smart Growth and allied principles, the desired elements of the EcoVillage include:

> new infill housing, built with ecological design features, a renovated neighborhood park, community gardens, and [a] rapid transit station, itself to reflect a green theme (incorporating passive and active solar features, a roof from recycled material, native landscaping), and within a short walk for residents of the neighborhood.
>
> (Wheeler and Beatley 2014: 424)

By 2004, twenty *new* "green" town homes had been built according to ecologically sensitive specifications with respect to solar energy, energy efficiency, and "green" construction materials (Wheeler and Beatley 2014). Inarguably, such building techniques, combined with the EcoVillage's emphases on compactness and alternative modes of transportation, contribute to environmental conservation. Yet, for all of their positives, ecovillages in general are frequently geared toward growth. The simple fact that such programs propose to construct *new* residential units in shrinking cities, where there are already vast oversupplies of housing (Ryan 2012), suggests that they are either attempts to attract people from outside their cities or are meant to catalyze intra-city "filtering" or "vacancy chain" processes (see Weaver and Holtkamp 2015 for an overview of these urban change models). In the former case, population growth is an overt objective, whereas the latter case contains the comparatively latent goal of fiscal (tax revenue) growth via gentrification-driven property value increases (Ch. 6). Regardless of which of these objectives applies, the underlying idea is patently growth-oriented. More acute examples of this *pro-growth* approach can be found in suburban or rural ecovillages – including the "Lakes of Orange" community just south of Cleveland – where converting greenfields into residential "green utopias" has been challenged as "greenwashing sprawl" (Schmitt 2012).

These examples of "sustainable development" in shrinking cities implicate a few broad points of contention over how sustainability is portrayed and pursued within the American model of urban development. Consider the three classic "values" of sustainability from an urban development perspective: ecology, economy, and equity (Godschalk 2004). Sustainable urban development is said to strike a balance between these values, such that economic growth occurs on the backdrop of ecological conservation and social equity. Clearly, tensions exist between these values. It has already been argued that American-style urban development produces socially uneven outcomes (a **property conflict** between economy and equity). In addition, economic activity necessarily expends natural capital (a **resource conflict** between ecology and economy). Finally, although economic growth can feasibly create jobs to "improve the lot of the poor", the varieties of low-skill jobs that tend to be suited to disadvantaged communities frequently involve ecologically destructive activities (Godschalk 2004: 6).

For that reason, programs of ecological conservation can disproportionately reduce employment opportunities for these communities (a **development conflict** between ecology and equity). Recently, a fourth value, *livability*, has been incorporated into this sustainability framework by Godschalk (2004). The addition of livability unearths at least three more conflicts: (1) a **growth management conflict** between livability and economy, whereby private transactions lead to market failures – investment is directed away from unprofitable spaces, where quality of life continues to decline; (2) a **green cities conflict** between livability and ecology, which concerns the degree to which human settlement patterns *ought to* manipulate the natural environment to enhance quality of life; and (3) a **gentrification conflict** between livability and equity, which concerns the displacement of existing residents as a result of quality-of-life improvements such as constructing ecovillages (Godschalk 2004).

When viewed through the lenses of the American model of urban development and the foregoing examples, many, especially Brundtland Report era instances of "sustainable development", are seen to favor the values of economy and ecology over equity and livability (e.g., Cook and Swyngedouw 2014). That is, pro-growth, yet environmentally sensitive, urban development projects like those just described for Hartford and Cleveland do not directly improve a city's capacity to create, test, and maintain adaptive capabilities in response to, say, social problems. It is feasible that economic growth can expand and diversify a city's economic portfolio, thereby making its economy – in the aggregate – more resilient to shocks such as deindustrialization. Moreover, building more ecologically sensitive spaces can improve those spaces' abilities to withstand environmental shocks like natural disasters. But, what is missing from consideration is how *individual agents* in cities are able to adapt and self-organize in the face of shocks. This is the true measure of a system's sustainability (Wu and Wu 2013), for a city's adaptive capacity depends on *how well its citizens can self-organize and adapt in response to disturbances* (Page 2011).

Consider again the general cases of a large-scale downtown Smart Growth "sustainable development", or a smaller-scale infill (ecovillage) "sustainable development", within a shrinking city. Now suppose, hypothetically, that both cases can be packaged as "green" tools for improving local conditions of vacancy and property blight (refer to Ch. 6–7). Despite their scalar and spatial differences, the conceptualization here is effectively the same: improvements to a given space will catalyze reinvestment in nearby spaces. However, even if this expectation bears out, and the development leads to localized gentrification processes, the policies do little or nothing to increase the ability of existing low-income households – who often occupy the neighborhoods most afflicted by vacancy and blight (Weaver and Bagchi-Sen 2014) – to partake in the positive changes and reinvestment. Rather, such households may be displaced by these developments, and with them conditions of vacancy and blight shift to another part of the city or region. This hypothetical situation uncovers an important point. Namely, sustainability is context dependent. Citywide increases in economic diversity or energy efficiency do not adequately capture the adaptive capabilities of a city's many, heterogeneous entities. Thus, urban *sustainability* goes well beyond the economic and environmental values described above; it involves enhancing the self-organization and adaptive capacities of individual agents.

In this sense, LEED buildings and insubstantial dependence on any one economic sector do not by themselves make cities "sustainable". Rather, these strategies merely reduce a city's *aggregate* exposure and/or sensitivity to stress (see above). Crucially, aggregate measures often suffer from significant loss of local information. What is needed for a city system to become sustainable, then, are policies that follow the third vulnerability-reducing strategy mentioned above: building adaptive capacity. This approach may take various forms, such as investing in human capital (health and education), strengthening democratic institutions, and so forth (Wu and Wu 2013). Regardless of the specific policy or institution though, building agent-level adaptive capacity in this manner is a necessary step toward improving livability and equity – the comparatively neglected "sustainability" values from urban planning – in city systems. The next wave of complex systems metaphors in urban planning and policy engages with these observations more directly.

Resilience planning

Modern resilience thinking in urban planning is sometimes viewed as a reaction to the destruction imposed on cities by recent high-profile natural disasters, such as Hurricane Katrina on the American Gulf Coast. In this sense, the discourse on "resilient cities" commonly concentrates on places' abilities to respond to external shocks that are brought about by nature (Newman, Beatley, and Boyer 2009). However, a growing community of scholars and practitioners is making deeper connections between cities and complex systems theory, and these contributions are expanding the purview of urban resilience to include disturbances of any kind, natural or otherwise (Pickett, Cadenasso, and Grove 2004; Eraydin and Taşan-Kok 2013; Pickett, Cadenasso, and McGrath 2013). This latter line of work is beginning to transform existing mental models associated with the pro-growth styles of "sustainability thinking" described above.

In particular, rather than emphasizing aggregate metrics or privileging the values of environment and economy over others (Cook and Swyngedouw 2014), resilience thinking starts at the bottom (focusing on livability and equity) and works its way up. More explicitly, resilience concerns how *vulnerabilities* are distributed socially, spatially, and temporally. Hence, as implied in the preceding section, insofar as exposure to and survival of shocks depends on *agent-level* attributes, the vast amount of heterogeneity within urban populations suggests that not all individuals are equally able to cope with disturbances. Consequently, no matter how ecologically sensitive or economically generative a particular "sustainable" urban development might be, if it does not enhance agents' adaptive capabilities – particularly for the most *vulnerable* agents – then it is unlikely to contribute to system-wide resilience, and, by extension, sustainability.

A large body of recent literature stresses this point. Much of this work is explicitly grounded in the concept of *resilience* (Renschler et al. 2010; Wu and Wu 2013), while other contributions exemplify resilience thinking under other headings, such as "just sustainabilities" (Agyeman 2005, 2013). Either way, the mental model is similar, and it departs markedly from the growth-oriented development logic outlined in Chapter 6 and revisited in the previous subsection. To be sure, recall that there are at least three generic strategies for improving a system's ability to "survive" stress: reduce its exposure to stress, decrease its sensitivity to stress, and strengthen its agents' capacities to adapt and self-organize. To the extent that "pro-growth sustainable development" (Zovanyi 2013) of the type just described diversifies overall urban economies and/or conserves [aggregate] environmental resources, such an approach plausibly reduces a city's *overall* exposure and sensitivity to specific economic and ecological stresses. However, further recall that equity and livability are important values in urban sustainability, and pro-growth development typically undermines these values, especially for the most vulnerable urban peoples and spaces (Godschalk 2004). With that said, the third strategy – enhancing agent-level adaptive capacities – can probably be viewed as the most efficacious way of creating resilient, sustainable urban systems (Agyeman 2005). This idea is a point of departure for contemporary resilience thinking relative to the first wave of "sustainable development" policy described above. Namely, the former is concerned with making cities more *sustainable* through strengthening their *resilience*, in line with the complex systems definitions introduced at the beginning of this chapter. In contrast, the latter was said to be concerned with achieving eco-technical notions of sustainability or *stability* on the foundation of [frequently acontextual] economic and population growth.

To make this contrast sharper, consider the following two examples. The first pertains to New Orleans, Louisiana, and its post-Hurricane Katrina recovery plan. Hurricane Katrina, the most expensive and one of the deadliest hurricanes in U.S. history (Knabb, Rhome, and Brown 2011), made landfall on the Louisiana coast on August 29, 2005 as a Category 3 storm. According to City of New Orleans officials, "[w]hile wind-related damages were extensive, it was the storm surge and subsequent flooding which caused [the city's] catastrophic level of loss" (City of New Orleans 2007: 25). The losses referred to in this statement were many, though one of the most prominent forms of loss was population contraction. Census Bureau estimates show that immediately prior to the storm, the city's population was approximately 465,000 residents (City of New Orleans 2007). Less than a year after the hurricane, this number had fallen to around 230,000 residents. In other words, the city's population was effectively halved (The Data Center 2014), which rapidly accelerated the (slower) shrinkage processes that were already taking place in New Orleans before the storm occurred (City of New Orleans 2007).

Much like the other shrinking cities discussed in this volume, this massive population loss was accompanied by spatially heterogeneous property abandonment and substantial physical decay in New Orleans' built environment. Although a large portion of the physical decay was the result of storm damage rather than property disinvestment *per se*, the presence of abandoned and derelict properties in a neighborhood nonetheless sends an antisocial signal, which makes endogenous neighborhood revitalization an unlikely outcome (Weaver and Holtkamp 2015). That being said, the City of New Orleans engaged in a post-Katrina participatory planning process that aimed to introduce strategic and geographically targeted shocks into the system in order to create conditions for future prosperity and neighborhood regeneration. The resultant Citywide Strategic Recovery and Rebuilding Plan (CSRRP) was underpinned by two main premises: (1) "everything [cannot be] fixed at once", meaning that incremental changes and small-scale innovations should be encouraged (see Marshall 2009); and (2) "it is unlikely that every former resident will return", suggesting that city officials ought to focus on improving quality of life for current and likely residents (City of New Orleans 2007: 9), rather than following popular pro-growth practices to increase population (Ch. 6).

These guiding principles, along with the city's commitments to public participation and promoting geographical evenness in socioeconomic and residential opportunities (City of New Orleans 2007), are emblematic of resilience planning. Indeed, the "strongest messages" that planners gleaned from public participation, which are echoed continuously throughout the CSRRP, can be directly tied to strategies for enhancing a system's resilience (Table 9.1; see City of New

Table 9.1 Community-identified priorities in New Orleans' Citywide Strategic Recovery and Rebuilding Plan (CSRRP)

Community Priority	*Description/Community Suggestions*	*Contribution(s) to Resilience*
Reduce Flood Risk	• Take a holistic approach to flood protection that includes wetlands restoration • Set voluntary standards for individuals to reduce their flood risk by making decisions to rebuild stronger or relocate safer	Reduce exposure to stress; increase adaptive capacity of residents to respond to stress
Empower Neighborhoods to Rebuild Safer and Stronger	• Provide financial incentives for residents to repair dilapidated real property • Provide information about spatially based risks to residents (as opposed to regulating where residents may or may not rebuild) • Offer incentives for neighbors to purchase blighted properties in their neighborhoods	Increase adaptive capacity of residents to respond to stress
Build Affordable Rental and Low-Income Housing	• Create affordable housing opportunities throughout the city (not spatially concentrated) to encourage spatial mobility	Reduce exposure; increase adaptive capacity
Reopen and Rebuild Public Facilities	• Ensure that all persons have access to public facilities such as schools and healthcare centers • Fully reopen and rebuild facilities in densely populated areas • Establish satellite or mobile facilities in less populated areas • Consolidate public services in multipurpose buildings for enhanced accessibility	Increase adaptive capacity of residents to respond to stress; decrease sensitivity to stress (access to reparative healthcare)
Rebuild Communities around High-Quality Schools	• Rebuild and reopen schools as 24/7 community centers • Improve quality of all schools	Increase adaptive capacity (human capital investment)

Source: City of New Orleans (2007: 53–54)

Orleans 2007: 53–54). Perhaps above all else, the CSRRP calls for and proposes mechanisms that "empower residents to rebuild . . . safe neighborhoods by providing financial incentives and the best possible information, rather than through government mandates and enforced standards" (City of New Orleans 2007: 54). Put differently, the recovery plan makes meaningful connections between agent-level adaptive capabilities and overall system (city) resilience and sustainability. Thus, in contrast to setting top-down goals for aggregate metrics such as the citywide vacancy rate,[1] the CSRRP seeks to facilitate bottom-up change through strategic neighborhood-level shocks. For instance, one action that the plan recommends is to offer incentives for neighbors – in contrast to developers or speculators – to acquire blighted properties in their neighborhoods (City of New Orleans 2007). Following from theories of neighborhood change, such a program enables local residents to remedy unsightly conditions in their communities, thereby contributing to a neighborhoodwide stabilization (Weaver and Holtkamp 2015). In this way, the recommended program strengthens neighbors' abilities to self-organize and adapt, which is a hallmark of system resilience (and, hence, sustainability).

It is possible to devote significantly more time and space to analyzing the New Orleans CSRRP. The plan recommends, contextualizes, and justifies specific, implementable policies and programs along thirteen dimensions: hurricane/flood protection, neighborhood stability, infrastructure and utilities, transportation, housing, the economy, healthcare, education, historic preservation/urban design, the environment, public safety, recreation and libraries, and other municipal and cultural facilities. However, it is beyond this chapter to address the particulars of a lengthy planning document. What is important here for practical purposes is the degree to which the plan's vision has manifested, and/or might manifest over time. Repeatedly, the CSRRP frames this vision as a "Safer, Stronger, and Smarter" New Orleans – a city that is "familiar, but different" (City of New Orleans 2007: 56). Translating this parsimonious, alliterative wish list of attributes into complex systems language, the specifics of the plan are effectively aimed at: reducing the city's exposure to disturbances (making it *safer*); decreasing the city's sensitivity to disturbances (making it *stronger*); increasing the capacity of the city's residents to respond and re-organize in the face of disturbances (making it[s residents] *smarter*); and ensuring that the city and all of its neighborhoods perform their essential functions in the changed, post-disturbance environment, regardless of whether they "look the same" as they did before the storm (making it *familiar, but different*). The overall vision of the CSRRP, then, is one of a *resilient*, and ultimately *sustainable*, city.

In the years since the plan was first drafted, the CSRRP vision has seemingly started to materialize and is continuing to take shape. For instance, the latest Census Bureau estimates show that the population of New Orleans is close to 379,000 residents.[2] This number represents more than four-fifths of the pre-Katrina population and an increase of roughly 65 percent over the initial post-Katrina estimate of 230,000 persons (The Data Center 2014). This level of population growth, which is certainly atypical of shrinking cities, plausibly reflects both the return of displaced residents,[3] and, consequently, the continued functioning of the city's diverse neighborhoods (e.g., City of New Orleans 2007). In addition, observers cite an "age of innovation" and "renaissance" in the coastal metropolis' economy (Editorial Board 2013). The president of the Rockefeller Foundation – the organization that pioneered a program called the "100 Resilient Cities" initiative[4] – has gone so far as to say that New Orleans should be nationally renowned as "The Resilient City" (Rodin 2014). Arguably, then, the CSRRP's internal focus on improving individual- and neighborhood-level quality of life through strategic interventions – e.g., empowerment programs, financial incentives, public service improvements, more and higher-quality public recreational and educational institutions, etc. – as opposed to an external focus on market-driven growth, has had at least some positive effects on the adaptive capacity of the city's residents.

Having said that, while New Orleans is still struggling with issues of vacancy, blight, economic disparities across races, and high crime rates (see endnote 3), such problems are not necessarily at odds with these perceptions of a resilient city. In fact, recall that the leading principle of the CSRRP is that "everything cannot be fixed at once". This perspective aligns squarely with the notion of evolution in complex adaptive systems, whereby incremental changes accumulate to produce systemwide consequences. The relatively short (evolutionary) time scale on which the recovery and rebuilding plan has operated to this point therefore suggests that "The Resilient City" is not a state that New Orleans boasts at present time, nor is it a static state of being at all. Instead, resilience (capacity) is an evolving systemic quality that can be built up or eroded over time. Where it appreciates, which seems to be happening in New Orleans, a city system moves toward *sustainability*. In this sense, the targeted interventions of the CSRRP might be operating in ways that will, over (evolutionary) time, make New Orleans a *sustainable city* that is characterized by high *resilience*.

New Orleans is not unique in its path toward *sustainability*. The elements of the New Orleans CSRRP read almost like a locally-adapted and context-sensitive version of the citywide "social urbanism" program initiated earlier in Medellín, Antioquia, Colombia.[5] This second example of resilience planning, a quick departure from the book's U.S.-centric approach, has been celebrated by urban planners and designers worldwide. It has even been highlighted as a model for American shrinking cities (Ryan 2012). Somewhat paradoxically, though, Medellín is not a shrinking city. To the contrary, it experienced population growth on the order of 620 percent during the same period (from 1950 onward) that American industrial centers were hollowing out (Ryan 2012). The fact that strategies adopted in a rapidly growing city might "work" for shrinking cities (e.g., New Orleans) therefore supports an important theme from earlier in this book: that growth and shrinkage are different outcomes of the same underlying processes (Beauregard 1993; Ch. 4). It follows that identifying and intervening at critical leverage points in the relevant systemic processes can plausibly "work" for vastly different cities, so long as the interventions are appropriate given the local context.

The *social urbanism* program in Medellín concentrated on a specific policy leverage point – poor urban neighborhoods – and introduced a variety of public works interventions. Just like many American shrinking cities, Medellín has historically been characterized by high poverty, property blight and dilapidated neighborhoods, high crime (especially violent and gang-related crime), substance abuse, and unemployment (Ryan 2012). Moreover, these problems tend to be spatially concentrated and co-occurring. In 2003, under newly elected, populist mayor Sergio Fajardo, city officials embarked on an aggressive "integrated urban project, to benefit poor neighborhoods comprehensively" (Ryan 2012: 175). Public transportation networks and infrastructure were built to connect poor, outer-ring communities to the city center; modern libraries with information technology and employment resources were set up in the most distressed areas, along with "quality schools"; existing schools were renovated and upgraded; local, neighborhood-based police stations were established to increase "eyes on the street"; and "enterprise centers" were sited in poorer neighborhoods to benefit low-income and unemployed individuals. What is more, all of these projects were carried out as prosocial acts of urban design: Fajardo sought to build "the most beautiful buildings in the poorest parts of the city" (Ryan 2012: 175–176).

The cumulative effects of the social urbanism project in Medellín have been remarkable. Like all cities, there is still crime, poverty, and unemployment. However, where Medellín was once one of the most violent places in the world, its homicide rate has dropped by more than 80 percent. To some observers, it has gone from "murder capital to model city".[6] Many attribute the city's successes, among them its selection to host the 2014 United Nations World Urban Forum [in part to showcase its progress], to continued public investments in the most underprivileged ("weak market") spaces and social groups in Medellín (see note 6).

By leveraging public works projects and public revenue to increase low-income individuals' geographic mobility, educational opportunities, access to recreational amenities, and security, the social urbanism program materially augments the *adaptive capabilities* of Medellín's citizens. This strategy is almost antithetical to the traditional American model of urban development (Ch. 6), in which underprivileged and distressed neighborhoods tend to repel private investment altogether, and where private investment is the primary mechanism for generating patterns of growth and decline in cities.

In this context, however, one might argue that New Orleans' post-Katrina recovery efforts have adopted a strategy somewhat closer to social urbanism than to the conventional American model. The result has been that both New Orleans and Medellín are now known for their *resilience* (Rodin 2014). Hence, despite the former city's characterization as "shrinking" and the latter's reputation as "growing" – as well as the former's embeddedness in American free-market democracy compared to the latter's presence in Colombian statism (Ryan 2012) – the common goal of increasing [especially low-income] residents' adaptive capabilities seems to be pushing the noticeably different systems toward similar, potentially *sustainable* (i.e., resilience-building, long-lasting) outcomes.

Concluding remarks

The extent to which a large-scale "social urbanism" project like the one undertaken in Medellín is feasible in the United States, given the political, economic, and legal frameworks present in American cities, remains to be seen (Ryan 2012), although the shift in demographic context towards multiethnicity – with new immigrants and new cultures – may be a reason for optimism (see Chapter 10). Nevertheless, this chapter has argued that *resilience* and *sustainability* strategies provide urban planners and researchers with a valuable apparatus on which to craft and test new, potentially "social urbanism-like" ideas for understanding and managing shrinking cities. In particular, sustainability is not, as it has been treated by adherents to the American model of urban development, simply about ecologically sensitive economic growth. Rather, to be *sustainable*, a city must be *resilient* – that is, it must adapt and self-organize to maintain its functions in the face of change. Moreover, it must be capable of becoming incrementally more resilient with time, in order to increase the range of shocks that it is capable of "surviving".

Three generic strategies exist to make a city more resilient over time (i.e., sustainable): (1) reduce its exposure to risk; (2) decrease its sensitivity to stress; and (3) increase the adaptive capabilities of its agents. Whereas an ideal policy program might pursue all three strategies using a variety of instruments (see Ch. 7), common "sustainable development" practices from the American model of urbanism (e.g., "signature" buildings and ecovillages) frequently identify with one or both of the former two but do not attend to the third of these. Crucially, though, it is the third strategy that plausibly contributes the most to resilience – for adaptation and self-organization occur at the level of individual citizens, not at the overall, aggregate system (Page 2011).

Where the *resilience* ideas have been embraced, such as in Medellín and New Orleans, the growth-first development logic of the neoliberal American model seems to be losing out to considerations of quality of life and equitable access to livable residential environments. This internal, palliative focus on improving the welfare of a place's existing residents, as opposed to an external penchant for attracting new residents, is functionally equivalent to increasing the adaptive capacity of a city system (Ryan 2012). More precisely, by providing new opportunities and resources to current citizens, especially those who live in historically disadvantaged and underprivileged communities, resilience planning programs effectively enhance the abilities of affected residents to cope with change. This attribute – unlike changes to building materials or neighborhood densities *per se* – is the crux of sustainability in (shrinking) cities. Toward that end, Medellín's "social urbanism" and New Orleans' CSRRP provide useful insights and first steps for transforming existing ways of thinking about urban development. The final chapter (Chapter 10) presents evidence of changing demographics in the United States and the need to contextualize further studies in shrinking cities within the spatio-temporal trends in population shifts, rising inequalities, and intergenerational opportunity.

Notes

1 See, for example, the discussion of the City of Buffalo "5 in 5" demolition plan in Chapter 6.
2 See: http://quickfacts.census.gov/qfd/states/22/2255000.html
3 See: http://www.cnn.com/2010/LIVING/08/29/katrina.new.orleans.resilient/
4 See: http://www.100resilientcities.org
5 This is an intentional overstatement. The two strategies are significantly different in specific content and political mechanisms; however, their attention to resilience make them comparable here.
6 See: http://www.theguardian.com/cities/2014/apr/17/medellin-murder-capital-to-model-city-miracle-un-world-urban-forum

References

Adger, W. Neil, and Katrina Brown. 2009. *Vulnerability and Resilience to Environmental Change: Ecological and Social Perspectives*. Oxford: Blackwell Publishing Ltd.

Agyeman, Julian. 2005. *Sustainable Communities and the Challenge of Environmental Justice*. New York, NY: New York University Press.

Agyeman, Julian. 2013. *Introducing Just Sustainabilities: Policy, Planning, and Practice*. New York: Zed books.

Agyeman, Julian, Robert D. Bullard, and Bob Evans. 2002. "Exploring the Nexus: Bringing Together Sustainability, Environmental Justice and Equity." *Space and Polity* 6 (1):77–90.

Batty, Michael. 2013. "A Theory of City Size." *Science* 340 (6139):1418–1419.

Beauregard, Robert A. 1993. "Representing Urban Decline Postwar Cities as Narrative Objects." *Urban Affairs Review* 29 (2):187–202.

Bulkeley, Harriet, and Michele Betsill. 2005. "Rethinking Sustainable Cities: Multilevel Governance and the 'Urban' Politics of Climate Change." *Environmental Politics* 14 (1):42–63.

City of Hartford. 2008. "Redevelopment Plan for the Constitution Plaza East Project." http://hartford.gov/economic-development/302-hra/431-redevelopment-and-neighborhood-services.

City of New Orleans. 2007. *Citywide Strategic Recovery and Rebuilding Plan: The Unified New Orleans Plan*. New Orleans, LA. http://resilience.abag.ca.gov/wp-content/documents/resilience/New%20Orleans-FINAL-PLAN-April-2007.pdf

Common, Michael, and Sigrid Stagl. 2005. *Ecological Economics: An Introduction*. Cambridge: Cambridge University Press.

Cook, Ian R., and Erik Swyngedouw. 2014. "Cities, Nature and Sustainability." In *Cities and Social Change*, edited by Ronan Paddison and Eugene McCann, 168–185. London: Sage.

de Vries, Bert J. M. 2013. *Sustainability Science*. New York: Cambridge University Press.

Editorial Board. 2013. "New Orleans Is an Exemplar of Resiliency: Editorial." *The Times-Picayune* December 13, 2013.

Eraydin, Ayda, and Tuna Taşan-Kok. 2013. "Introduction: Resilience Thinking in Urban Planning." In *Resilience Thinking in Urban Planning*, edited by Ayda Eraydin and Tuna Taşan-Kok, 1–16. New York: Springer.

Godschalk, David R. 2004. "Land Use Planning Challenges: Coping with Conflicts in Visions of Sustainable Development and Livable Communities." *Journal of the American Planning Association* 70 (1):5–13.

Greenberg, Miriam. 2013. "What on Earth Is Sustainable? Toward Critical Sustainability Studies." *Boom: A Journal of California* 3 (4):12.

Holling, Crawford S. 1973. "Resilience and Stability of Ecological Systems." *Annual Review of Ecology and Systematics* 4:1–23.

Holling, Crawford S. 2001. "Understanding the Complexity of Economic, Ecological, and Social Systems." *Ecosystems* 4 (5):390–405.

Hospers, Gert-Jan. 2014. "Urban Shrinkage in the EU." In *Shrinking Cities: A Global Perspective*, edited by Harry W. Richardson and Chang Woon Nam 47–58. London: Routledge.

Knabb, R.D., J.R. Rhome, and D.P. Brown. 2011. National Hurricane Center. Tropical Cyclone Report: Hurricane Katrina, 23–30 August 2005.

LaCroix, Catherine J. 2011. "Urban Green Uses: The New Renewal." *Planning & Environmental Law* 63 (5):3–13.

Lewis, Linda. 2000. "Environmental Audits in Local Government: A Useful Means to Progress in Sustainable Development." *Accounting Forum* 24 (3):296–318.

Mansfield, Becky. 2009. "Sustainability." In *A Companion to Environmental Geography*, edited by Noel Castree, David Demeritt, Diana Liverman and Bruce Rhoads, 37–49. London: Blackwell.

Marshall, Stephen. 2009. *Cities Design & Evolution*. London, UK: Routledge.

Newman, Peter, Timothy Beatley, and Heather M. Boyer. 2009. *Resilient Cities: Responding to Peak Oil and Climate Change*. 1st ed. Washington, DC: Island Press.

Page, Scott E. 2011. *Diversity and Complexity*. Princeton, NJ: Princeton University Press.

Pickett, S. T. A., M. L. Cadenasso, and Brian McGrath, eds. 2013. *Resilience in Ecology and Urban Design Linking Theory and Practice for Sustainable Cities*. New York: Springer.

Pickett, Steward T.A., Mary L. Cadenasso, and J. Morgan Grove. 2004. "Resilient Cities: Meaning, Models, and Metaphor for Integrating the Ecological, Socio-Economic, and Planning Realms." *Landscape and Urban Planning* 69 (4):369–384.

Platt, Rutherford H. 2014. *Land Use and Society: Geography, Law, and Public Policy*. 2nd ed. Washington, DC: Island Press/Center for Resource Economics.

Renschler, Chris S., Amy E. Fraizer, Lucy A. Arendt, Gian-Paolo Cimellaro, Andrei M. Reinhorn, and Michel Bruneau. 2010. *A Framework for Defining and Measuring Resilience at the Community Scale: The Peoples Resilience Framework*. Buffalo: MCEER.

Rodin, Judith. 2014. "New Orleans Should Nickname Itself the Resilient City." *The Times-Picayune*, November 07, 2014.

Ryan, Brent D. 2012. *Design after Decline: How America Rebuilds Shrinking Cities*. Philadelphia: University of Pennsylvania Press.

Schilling, Joseph, and Jonathan Logan. 2008. "Greening the Rust Belt: A Green Infrastructure Model for Right Sizing America's Shrinking Cities." *Journal of the American Planning Association* 74 (4):451–466.

Schilling, Joseph M., and Alan Mallach. 2012. *Cities in Transition: A Guide for Practicing Planners*. Vol. 568, *Planning Advisory Service*. Chicago, IL: American Planning Association.

Schmitt, Angie. 2012. "The Greenwashing of Sprawl." STREETSBLOG USA. http://usa.streetsblog.org/2012/04/11/the-greenwashing-of-sprawl/.

Scott, M. Robert. 2002. "Gettin' Easier Bein' Green: Eco-City Thrives in Cleveland." *Ohio Realtor*, September, 2002.

Taşan-Kok, Tuna, Dominic Stead, and Peiwen Lu. 2013. "Conceptual Overview of Resilience: History and Context." In *Resilience Thinking in Urban Planning*, edited by Ayda Eraydin and Tuna Taşan-Kok, 39–51. New York: Springer.

The Data Center, "Facts for Features: Katrina Impact," August 12, 2015, available at: http://www.datacenterresearch.org/data-resources/katrina/facts-for-impact/

Voinov, Alexey. 2008. "Understanding and Communicating Sustainability: Global Versus Regional Perspectives." *Environment, Development and Sustainability* 10 (4):487–501.

Warner, Daniel M. 2006. "Commentary: 'Post-Growthism': From Smart Growth to Sustainable Development." *Environmental Practice* 8 (03):169–179.

Weaver, Russell, and Chris Holtkamp. 2015. "Geographical Approaches to Understanding Urban Decline: From Evolutionary Theory to Political Economy . . . and Back?" *Geography Compass* 9 (5):286–302.

Weaver, Russell C., and Sharmistha Bagchi-Sen. 2014. "Evolutionary Analysis of Neighborhood Decline Using Multilevel Selection Theory." *Annals of the Association of American Geographers* 104 (4):765–783.

Wheeler, Stephen M., and Timothy Beatley. 2014. *Sustainable Urban Development Reader*. New York: Routledge.

World Commission on Environment and Development. 1987. *Our Common Future*. Oxford: Oxford University Press.

Wu, Jianguo, and Tong Wu. 2013. "Ecological Resilience as a Foundation for Urban Design and Sustainability." In *Resilience in Ecology and Urban Design*, edited by S. T. A. Pickett, M. L. Cadenasso and Brian McGrath, 211–229. New York, NY: Springer.

Zovanyi, Gabor. 2013. *The No-Growth Imperative: Creating Sustainable Communities under Ecological Limits to Growth*. New York: Routledge.

10 Concluding remarks

> In addition to the prevalence and severity of population loss, we need to assess its persistence; that is, the extent to which individual cities shed residents from one decade to the next. A population drop confined to a single decade is a temporary set-back; losing residents over two decades is cause for alarm. The latter indicates a more daunting problem for residents, investors, and public officials. Moreover, it hints at a structural rather than a circumstantial impediment to the city's ability to grow.
>
> (Beauregard 2009: 64)

In this book, we first offered empirical analysis to answer theoretically grounded questions and illustrate patterns of shrinkage and decline in the United States (Chs. 2 through 5). We then discussed policies and governance patterns in Chapters 6, 7 and 8, followed by a discussion of resilience and sustainability in Chapter 9. The empirical analyses show that shrinkage has spread geographically beyond the Rust Belt and locally beyond central cities (Chs. 4 and 5). However, these patterns vary widely, and their complexity highlights the need to understand the interaction of social factors with the three main drivers of shrinkage: demographic change, deindustrialization, and suburbanization. This complexity also draws attention to the emergence of alternative ways of thinking about urban development such as rightsizing, which is a challenge to existing pro-growth approaches in the American model of urban development. In recent years, conceptual thinking about sustainability and resilience has been operationalized to address these complicated place-based problems (e.g., post-Katrina New Orleans). While this book hypothesized and tested several questions surrounding shrinkage and decline, the empirical analyses also raised more questions about the manifestation of shrinkage. This chapter will discuss some of the findings and highlight additional trends observed in the United States relevant to scholars and practitioners interested in researching intra-urban trends in shrinking cities vis-à-vis other cities.

Summary of key findings

Population loss is considered to be the central issue in understanding the geographic patterns of shrinkage. Although Rust Belt cities have been the main sufferers, the 1970–2010 data analysis shows the emergence of shrinking census tracts in otherwise non-shrinking cities in other parts of the United States beyond the Rust Belt. Interwoven with population shrinkage is economic shrinkage, and the constantly evolving relationship between the two – or continuous downturn – is identified as negative cumulative causation. Although shrinking places (i.e., census tracts) can be found in all types of growing, shrinking, and stable cities, disadvantage – measured using the share of non-white population, female-headed household, unemployed, poverty, and low education – is far more acute in shrinking tracts. Following on the causes and effects of population loss, a major *outcome* of shrinkage is the increase in vacant, abandoned, and tax-delinquent residential and non-residential buildings. In other words, persistent population shrinkage is also deeply connected with decline in the built environment. However, shrinking tracts in cities and suburbs that withstood pressures to undergo severe decline were found to have higher levels of social capital – measured by stable homeownership rates, a relatively high number of civic organizations, a relatively high number of government offices, a greater share of long-term stakeholders, a greater predisposition to trust others in transactions characterized by asymmetric information, slightly lower income inequality, and greater internal community homogeneity.

The first part of the book (Chs. 2–5), therefore, shows the regionalization and suburbanization of shrinkage along with a discussion of the importance of social capital in differentiating among different categories of shrinkage/decline. Specific findings show that (i) 66.1 percent of census tracts that have been shrinking since 2000 are within central cities compared to 69.5 percent of tracts that shrank between 1970 and 2010; (ii) 56.5 percent of tracts that declined between 2000 and 2010 are within central cities compared to 69.5 percent of tracts that declined from 1970 to 2010; and (iii) 72.0 percent of tracts that shrank and declined between 2000 and 2010 are within central cities compared to 83.7 percent of tracts that shrank and declined from 1970 to 2010. The key finding here is that the pattern of shrinkage and decline is rapidly evolving in non-central city tracts.

The second half of the book (Chs. 6–9) offered a discussion of pro-growth policies and alternatives to these typical "American model" approaches such as rightsizing and smart decline followed by a discussion of governance and approaches to urban resilience and sustainability practices. These chapters focused on (i) weaknesses with pro-growth approaches to urban development, (ii) difficulties in implementing and accepting the rightsizing and smart decline approach, (iii) complexity of governance and difficulties in defragmenting governance – there is no one-size-fits-all formula, and (iv) scalar issues, among others, in implementing sustainability and resilience practices. In other words, what works in one area (e.g., census tract) may not be relevant for the whole county or the city (e.g., modifiable areal unit problem) – practices from one part of the United States to another need not work without a detailed understanding of the demographic, socio-economic, and political contexts.

Furthermore, answers to questions are limited by the type of data available and the tools of analysis. While the tools have become quite sophisticated, variables are limited to those factors that can be measured, and certain variables are not available for all types of geographic regions. Therefore, systematic case studies of existing and emerging cities and regions, with various combinations of shrinking and declining tracts, are necessary to provide an understanding of processes, policies, and outcomes so that communities can learn from each other's best practices and mistakes. This book drew upon many examples, but historic analyses of representative cases are beyond our scope. Below, some areas of future research are discussed with a specific focus on demographic shifts in the United States and the implications for urban research. A brief discussion of spatial analysis is added at the end to emphasize the need for extensive data analysis, which can detect emerging patterns and direct us to do in-depth comparative urban research to inform stakeholders.

Emerging debates

One of the issues arising in the United States is the so-called "Great Inversion" whereby young professionals and affluent middle-aged residents are moving back into cities and immigrants and lower-income residents are moving out to the suburbs (Ehrenhalt 2012). The narrative of cities as places of crime, disorder, and poverty is being reshaped in America, and some have suggested a return to prominence (Glaeser 2011; Ehrenhalt 2012). However, many of these narratives miss how recent pro-growth policies have "uncoupled" the city into two divergent parts: the "economic city" and the "demographic city" (Mallach 2015). The patterns of reinvestment have always been uneven, but in the last decade, that pattern has been one of growth in jobs (typically highly subsidized) and wealth in central business districts while poverty and population decline continue in many neighborhoods throughout these same cities (Mallach 2015). So in places like Cleveland, where population declined by 17 percent in the 2000s, the downtown population has recently seen an uptick, fueled by pro-growth strategies that privilege the "economic city" over the "demographic city." This dichotomy calls for understanding the demographic shifts and economic shifts in the United States in order to contextualize city-level issues. In the following section, we provide some demographic data to show the need to examine all components of demographic shift to understand who will make up the labor force of the city in the near future and in the long term. Further, we draw attention to economic shifts (e.g., among others, two topics of importance are intra-urban inequality and inter-generational mobility) and the need to couple economic and demographic analysis for place-based policy development and planning.

Demographic trends

In a brief article discussing the recent U.S. trend of rising natural decrease, that is, deaths exceeding births (e.g., over 2.5 million deaths in 2012 exceeded numbers from previous years), Johnson (2013a: 2) notes "Demography is not destiny, but one ignores it at their peril". Often thought of as a rural phenomenon, county-level data analyses show that urban counties are experiencing natural decrease as well. Natural decrease is a function of the local age structure (e.g., adults of child-bearing age) and the proportion of older adults who are at a high risk of mortality. Regions losing young adults experience aging in place, but at the same time, retirement destinations in parts of Florida, Arizona, the Upper Great Lakes, and New England, for example, have seen a rise in population with a high risk of mortality. Findings such as these highlight the exportation of population problems from one region to another – that is, soon these states will face shrinkage and may have to manage infrastructure to fit their existing population.

Figure 10.1, produced by the Carolina Population Center, shows the patterns of the components of population change in the United States in 2010–2014 (Tippett 2015a). The results are central to stakeholders interested in contextualizing short- and long-term trends within overall regional and U.S.-wide patterns to better understand their risk for experiencing shrinkage and decline. Only 13 percent of all counties noted population growth from natural increase alone, while 6.5 percent experienced growth from net in-migration alone. Only 28 percent of all counties in the United States grew based on both natural increase and net migration from 2010–2014, while 24 percent (754 counties) lost population from these two factors combined (net out-migration and natural decrease). Exclusive net out-migration affected another 758 counties while natural decrease affected only 5 percent of all U.S. counties.

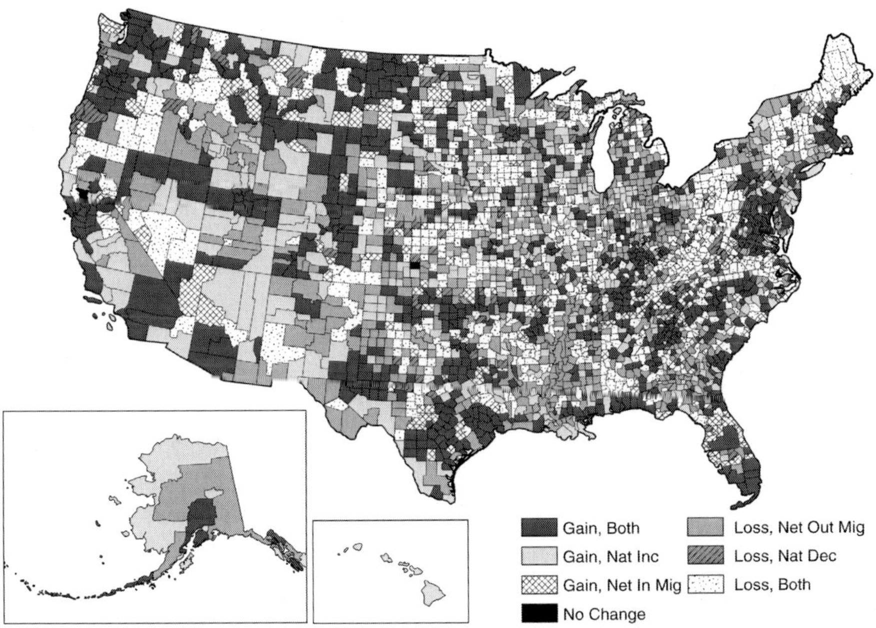

Figure 10.1 Components of population change in the United States (Tippett 2015a: used with permission from the Carolina Population Center)

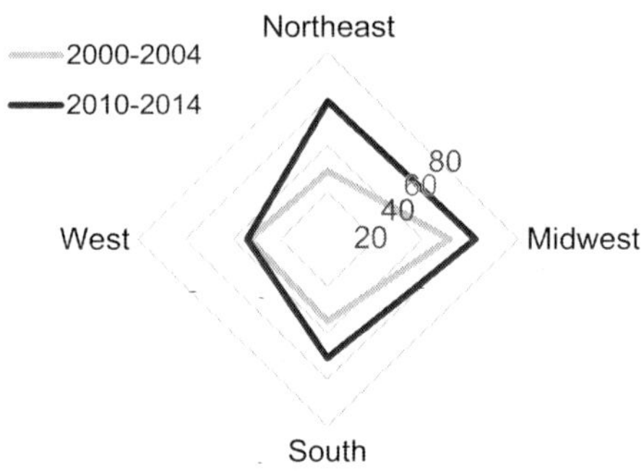

Figure 10.2 Share of counties with population loss (%) by region in the United States (Data source: Tippett 2015a)

These findings are important for shrinkage and decline because in recent years, net out-migration has actually been the major component of population loss in counties:

> A decade ago, the Midwest was the only region where the majority of counties lost population between 2000 and 2004. In 2014, it maintained its position as the region with the largest share of counties experiencing population loss since the 2010 census: 62 percent. Two other regions, the Northeast and South, also had the majority of their counties lose population between 2010 and 2014.
> (Tippett 2015a: 1)

Tippet's observation is illustrated in Figure 10.2. Furthermore, between 2010 and 2014, 91 percent of all counties that lost population experienced net out-migration, and 54 percent of counties experienced natural decrease: ". . . migration was the key factor in county population losses nationwide" (Tippett 2015b: 1).

Given that the central focus in understanding shrinking cities is population change, deeper analysis of the demographic components and the selection of various range in terms of time periods can provide a backdrop (usually available at the county level) within which to analyze related socio-economic changes at finer geographic scales (e.g., census tracts or block groups). For example, Erie County in western New York is, as a whole, not shrinking, but the City of Buffalo, nested within Erie County, is not only shrinking but certain parts are undergoing decline. A finer, contextual analysis of Buffalo shows patterns of actually existing shrinkage through an *uncoupling* of the economic city from the demographic city, as substantial inflows of capital investment (highly subsidized) flood the central business district and a growing medical corridor while little investment and no population growth is happening in many neighborhoods. Furthermore, the broader context of the Northeast shows overall loss (Figure 10.2), with long-term funding issues across the states.

Adding to the complexity of demographic change, many city tracts have been gaining foreign-born population in recent decades (Fig. 10.3). These trends are not

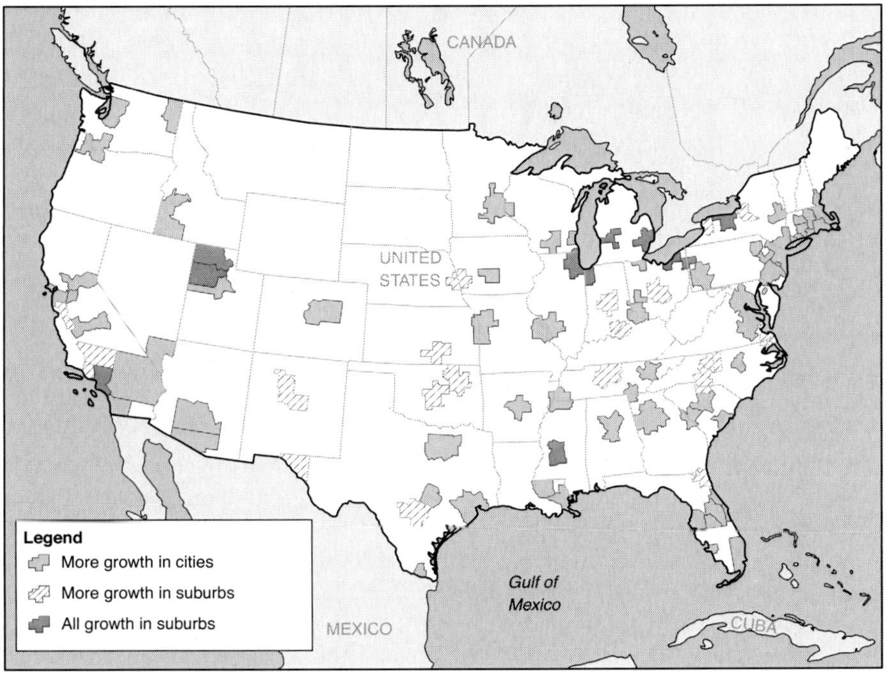

Figure 10.3 Foreign-born population growth in cities and suburbs, by core-based statistical area (CBSA), 2000–2013 (Data source: Wilson and Svajlenka 2014)

uniform across cities or even regionalized, though (e.g., the City of Buffalo has been gaining population due to foreign-born residents, but the City of Cleveland has not). And it should be noted that not a single metro area saw its foreign-born population in the suburbs decrease between 2000 and 2013. In twenty metros, the suburban immigrant population at least doubled (Fig. 10.3). Therefore, scholarship and policymaking need to focus on the details of place-based change contextualized within different governance areas to be effective in terms of rightsizing, smart decline, sustainability, and/or resilience.

Several caveats need to be addressed (Airriess 2016) with respect to positive population change with a specific focus on new immigrant migrations. First, new foreign-born immigrants tend to cluster in gateway cities (e.g., Boston, Los Angeles, etc.), but there are new patterns emerging in mid-size cities (e.g., a large Burmese refugee inflow in Buffalo). Second, immigrants with higher levels of education (e.g., South and East Asians) tend to make suburbs as their residential choice and therefore do not offset the overall loss in either human or social capital (Skop 2016). Third, a fundamental issue for immigrants arriving in a new city is economic opportunity (see the case of Poughkeepsie outlined below) – even if in-migration is able to offset the population loss, economic opportunities are not always abundant, which means economic gains are rarely spectacular. Finally, there is the propensity of some immigrant groups to pursue the "American Dream" and move quickly out to the suburbs without ever buying property (hence investing) in the city (Airriess 2016). In sum, there is a critical need to contextualize discussions of migrants or immigrants and their role in overall development of a place (e.g., human and social capital) within the broader demographic trends of the city and surrounding metropolitan area so as not to mask intra-regional variations or city-to-city variations. For example, if "non-metropolitan" is added to the mix, foreign-born components need to be examined given the increasing contribution of this group to population and socio-cultural change (see Johnson 2013b).

Relatedly, in addition to the *number* of new immigrants moving to shrinking and other cities, the *diversity* of those immigrants in terms of race and ethnicity is important for understanding the emerging settlement patterns in U.S. cities. The two major groups of foreign-born populations in the 21st century United States have been Latin and Caribbean American ethnicities and Asians and Middle Eastern ethnicities. The subgroups within these two major categories show diverse settlement patterns in the United States thereby altering cultural landscapes, which has long-term implications for the local-regional economy. The book *Contemporary Ethnic Geographies in America* (Airriess 2016) tackles various categories of immigrant ethnic groups, both legal and illegal, and their imprint on America's cultural landscapes. The number of foreign born people living in the United States increased from 27.6 million in 2000 to 39.9 million in 2010 – 53 and 25 percent were from Latin America and Asia, respectively, with Mexicans making up the largest percentage (55 percent) of Latin Americans. Similarly, the domination of ethnic Chinese (17 percent) is seeing a change toward Asian Indians (16 percent). By 2050, the proportion of the population that is of non-European descent is expected to increase to 47 percent from 31 percent in 2000. By 2060, the foreign-born population in the United States will likely be nearly 19 percent (see Box10.1).

Regionally, the new immigrant geographies (as measured by the share of foreign-born population) show that between 1980 and 2010, the South and West regions increased their share of foreign-born residents in contrast to the traditional destinations of the Northeast and the Midwest. The South's share went up from 20.6 to 31.9 percent, the West's share went up from 32.4 to 35.3 percent, and the share of the Northeast and the Midwest declined from 32 percent to 21.6 percent and 15 to 11.2 percent, respectively. While the traditional gateways (e.g., New York, Philadelphia, Chicago, Los Angeles, and San Francisco) still receive immigrants, new gateways have emerged, with four metropolitan areas recording unprecedented growth: Charlotte, North Carolina; Atlanta, Georgia; Austin, Texas; and Las Vegas, Nevada (Airriess 2016). Contrasting the conditions in these cities with typical shrinking city types is the next step toward understanding the emergent relationships between population change and income/wealth generation. In other words, if shrinking cities can stabilize, are they capable of offering better prospects (e.g., inter-generational mobility) to new migrants?

In contrast to historical patterns of migration, emerging trends show that the first residential destination for new immigrants is often the suburb.

Box 10.1 Demographic shifts in U.S. population

According to the report, *Projections of the Size and Composition of the U.S. Population: 2014 to 2060*:

- The U.S. population is expected to grow more slowly in future decades than it did in the previous century. Nonetheless, the total population of 319 million in 2014 is projected to reach the 400 million threshold in 2051 and 417 million in 2060.
- Around the time the 2020 Census is conducted, more than half of the nation's children are expected to be part of a minority race or ethnic group. This proportion is expected to continue to grow so that by 2060, just 36 percent of all children (people under age 18) will be single-race non-Hispanic white, compared with 52 percent today.
- The U.S. population as a whole is expected to follow a similar trend, becoming majority-minority in 2044. The minority population is projected to rise to 56 percent of the total in 2060, compared with 38 percent in 2014.
- While one milestone would be reached by the 2020 Census, another will be achieved by the 2030 Census: all baby boomers will have reached age 65 or older (this will actually occur in 2029). Consequently, in that year, one-in-five Americans would be 65 or older, up from one in seven in 2014.
- By 2060, the nation's foreign-born population would reach nearly 19 percent of the total population, up from 13 percent in 2014

Source (quoted from): http://www.census.gov/newsroom/press-releases/2015/cb15-tps16.html accessed January 22, 2016

For example, in the 97 largest metropolitan areas of the United States, the number of immigrants living in the suburbs increased from 56 percent in 2000 to 61 percent in 2013 (Note: 69 percent of all residents in these large metro areas live in suburbs – therefore, immigrants are slightly under-represented in the suburbs) (see Wilson and Svajlenka 2014, also cited in Airriess 2016). A summary of findings of immigrant distribution (Wilson and Svajlenka 2014: 1) is as follows:

- "Immigrants continue to be attracted to the nation's largest metropolitan areas but are dispersing to more and smaller places across the country" (e.g., Baltimore, Maryland; Charlotte, North Carolina; Minneapolis, Minnesota).
- "The New York metropolitan area . . . saw the greatest increase in its foreign-born population between 2000 and 2013: almost 800,000."
- "Eighty seven of the 100 largest metro areas' immigrants made up a larger share of the population in 2013 than they did in 2000."

- "While large immigrant gateways saw the highest growth in their foreign-born populations over the 13-year period, smaller places experienced very rapid growth" (e.g., Scranton, Pennsylvania; Cape Coral, Florida; Knoxville, Tennessee; Indianapolis, Indiana; Nashville, Tennessee; Charlotte, North Carolina; Louisville, Kentucky; Charleston, South Carolina; and Raleigh, North Carolina)
- "As immigrants continued to disperse to more metro areas around the country, they also continued to move to the suburbs, following overall population settlement patterns"

Only twelve metro areas noted more immigrants coming to the central city compared to their suburbs – Charlotte, North Carolina; Colorado Springs, Colorado; Columbus, Ohio; El Paso, Texas; Jacksonville, Florida; Nashville, Tennessee; Oklahoma City, Oklahoma; Omaha, Nebraska; San Antonio, Texas; San Jose, California; Tulsa, Oklahoma; and Wichita, Kansas (see Figure 10.3). Except for Columbus, Ohio, no other cities can compete with their suburbs in the Northeast and the Midwest for foreign-born migrants. As such, comparative urban research is needed to understand the trends in socio-economic development of immigrants, who are often ethnically diverse.

Whether place-making by new ethnic groups is a way forward remains to be seen, although many examples are given in Airriess (2016). Stability of these communities, patterns of income distribution, and the emergence of various forms of social capital are topics that should be explored in addition to understanding the displacement of the poor from some neighborhoods in shrinking cities that are desperate to revitalize. Godfrey (2016) offers three case studies of ethnic place-making to highlight the following types of urban ethnic landscapes: incipient landscapes of ethnic arrival, consolidated landscapes of ethnic enclaves, and defensive landscapes of ethnic contention. In order to understand what eventually affects a city, an understanding of neighborhood-level transitions is critical.

Poughkeepsie is a classic example of ethnic place-making. In particular, the transition of the city's Middle Main Street neighborhood from white and black to Hispanic (44 percent of 2010 population as compared to 9.7 percent in 1990) exemplifies incipient landscapes of ethnic arrival. In contrast, the case for New York's Loisaida or East Village or San Francisco's Inner Mission districts show defensive landscapes where there has been a decline in percent black and Hispanic and an increase in white and Asian population. Chinatown in San Francisco and Harlem in New York exemplify consolidation of ethnic-place-based identities. In these two districts, Asians (Chinatown) and African Americans (Harlem) continue their stronghold although some displacement has been noticed due to redevelopment.

Introduced above, Poughkeepsie is a classic example of a shrinking city that regained a substantial amount of its population from Latino influx. The population in 2010 of 32,736 is still below the peak of 41,000 in 1950, but chain migration, particularly from Oaxaca, Mexico, has been transforming the central city, which underwent widespread demolition of about 4,000 units in the post-war urban renewal era. The immigrants initially joined low-wage jobs, but in recent years ethnic entrepreneurship can be noted, especially in the retail business. Although this transformation cannot be labeled as gentrification given that 30 percent of the population continues to live at or below the poverty level, the 2010 median income still doubled over the 2000 figure, indicating gradual improvement of the economy.

In comparison, the situation in Loisaida, New York, shows the influx of affluent yet mobile groups (e.g., short-term renters such as students enrolled at New York University). Anti-displacement initiatives have actually improved media coverage and have made the place attractive to gentrification (see Godfrey 2016 for detailed case histories of these districts undergoing change in the recent decades). The main point of these few examples is that the ethnic components of in-migration, especially that of foreign-born population, need to be considered in future research.

Economic trends

Related topics of interest for place-based analysis are inequality and inter-generational mobility. The attention is now shifting from global measures to local manifestation of inequalities with a focus toward inter-generational mobility – leaving researchers to questions whether conditions exist to help the children of new immigrants, or whether they will be entrenched in poverty (Airriess 2016). Berube (2014) notes that big cities have huge income gaps between households in the 95th percentile of income distribution compared to those in the 20th percentile of income distribution. Rogerson (2015) offers a method to compare whether inequality within low-income groups is higher than the inequality within high-income groups in a geographic unit. In other words, there are various ways in which we can examine inequality. However, an understanding of the place-based characteristics that create these inequalities is critical for halting the continued downward spiral toward the conditions characteristic of many developing countries in certain parts of U.S. cities. As shown below (and in Ch. 5), these conditions are often not contained in cities and have spillover effects to the suburbs.

As Berube (2014) notes, unequal cities are not created equally – creating implications for policy mobility from one city to another and also the intra-metropolitan practices of various stakeholders. For instance, Miami and San Francisco show similar ratios of household income between the 95th percentile and the 20th percentile. However, while San Francisco's wealthiest residents boast very high incomes, Miami, in contrast, has many poor residents with few enclaves of high-income earners. Cities with low levels of income inequality are mostly Southern and Western cities – their changing inequalities were affected by the recession (see Appendix C) – similarly, Midwestern cities show inequality issues following additional manufacturing decline in the recession period. The 2007–2012 shifts in inequality show that the rich are not necessarily getting richer, but the poor are definitely getting poorer (e.g., Atlanta's average household income at the 20th percentile was $14,850, but the loss was $4,036 between 2007 and 2012). The range of income in cities with the highest inequality (Appendix C) at the 20th percentile was $10,438 in Miami (which lost $1,840 between 2007 and 2012) to $21,782 in

Washington, D.C. While shrinking and declining cities are not being discussed, their historic and current patterns may bring attention to cultures and norms that worked or did not work. Furthermore, such analyses may show that population increase in shrinking cities may not lead to economic turnaround.

The relationship between income inequality and wage inequality is not straightforward. However, income inequalities in cities have triggered debates about the minimum wage. One important point is that a small shift in income inequality in the city may not just mean that the city has stable conditions but may reflect suburbanization of the poor. With very little disaggregated analysis by groups and place (e.g., census tracts or blocks) contextualized within the broader demographic trends in a city versus its metropolitan region or suburbs, many issues such as cohort and spillover effects (see discussion below) are missed.

The topic of inequality and inter-generational mobility is a critical starting point for demonstrating the importance of conducting comparative urban research. Are the shrinking cities of the United States ready to absorb a large influx of population from various ethnic groups all around the world? Although authors often caution against putting too much emphasis on causality, questions are raised about geographic variations in inter-generational mobility and hence the future of places and people within U.S. cities. Chetty et al. (2014) use federal income tax record data (1996–2012) to study the geography of inter-generational mobility. Their study finds five factors to be strongly associated with inter-generational mobility: (1) high residential segregation is correlated with less mobility; (2) areas with more inequality as measured by Gini coefficients are correlated with less mobility; (3) the quality of the K–12 school system is negatively correlated with mobility; (4) social capital indices help mobility; and (5) weaker family structures (e.g., single-parent households) are associated with less mobility. While the study has many conclusions, a key point is that factors hurting the middle class may hurt inter-generational mobility more than changes (positive) in income for those at the highest level of income distribution.

In another study, Chetty, Hendren, and Katz (2015) discuss whether children under 13 benefited from the Moving to Opportunity for Fair Housing program of the U.S. Department of Housing and Urban Development (HUD). Although previous studies did not find any significant findings on earnings and employment of adults, the current study shows that neighborhood effects are significant in terms of developmental effects of children under 13:

> More broadly, our findings suggest that efforts to integrate disadvantaged families into mixed-income communities are likely to reduce the persistence of poverty across generations.
>
> (Chetty, Hendren, and Katz 2015: 41)

The authors argue that inter-generational mobility issues are a local problem, and therefore, place-based policies are critical. Sawhill (2015) states:

> We should invest in evidence-based programs, starting before birth and extending through high school and the college years. If we provided an effective

home visiting program, high quality pre-k, and comprehensive school reforms in elementary and high school, it would make a difference in children's lives, according to rigorous experimental evidence.

However, what is not fully understood is "why some areas of the United States generate higher rates of mobility than others" (Chetty et al. 2014: 42). Local structures and opportunities interacting with demographic factors (e.g., components of population loss/gain) need to be examined using new data sources and analytical techniques. Many topics can be examined: Are shrinking cities experiencing better or worse inter-generational mobility compared to other groups? How are the patterns different between natives and foreign-born? The following section provides a quick overview of some of the techniques that may be useful in detecting patterns, if data are available.

Detecting patterns of intra-regional variation

Cluster detection

Throughout the book, particularly in Chapters 2, 3, and 4, we used several different types of statistical tests to draw empirical evidence regarding shrinkage and decline. While these statistics are useful for comparing conditions of population loss and decline between cities in the United States, finer-scale analyses are often needed to understand the variabilities discussed above, such as local manifestations of immigrant mobility or income inequalities. While an in-depth treatment of the many available statistical measures is beyond the scope of this book, a brief discussion will show possibilities for future research to monitor emerging spatial patterns over time.

Spatial clustering techniques offer a means for quantifying uneven geographic variation of a particular variable or variables. Cluster detection techniques are based on the premise of **spatial autocorrelation**, which is a measure of the degree to which features tend to be clustered together or dispersed through space. Features that are positively spatially autocorrelated are clustered together in space while features that are negatively autocorrelated are dispersed. For example, in a shrinking city, we might expect a declining neighborhood to be next to, or at least in close proximity to, other declining neighborhoods (i.e., they are positively spatially autocorrelated). In contrast, we would not expect to find a checkerboard pattern of declining neighborhoods interspersed with thriving neighborhoods all throughout the entire shrinking city. Stated otherwise, things that are alike tend to be located near to each other. When focusing on the uneven patterns of development in cities, such metrics can highlight where within the city the patterns are non-random, which in turn can reveal information about the underlying spatial processes that might be producing those variations.

Within these cluster detection techniques, there are two broad categories of measures: global and local. Global measures provide an overall indication of whether spatial interdependence exists across the study area through a single value. Some of the most commonly used global clustering techniques include Moran's I (Moran 1950), Geary's C, and Getis-Ord General G (Getis and Ord 1992) statistics. The second category, local measures, tests for spatial association of a variable within a geographic neighborhood. Some commonly used local autocorrelation measures are the Local Indicators of Spatial Association (LISA; Anselin 1996) and the Getis-Ord Gi* statistic (Ord and Getis 1995). Both of these statistics can identify local patterns of spatial association, which are commonly known as clusters or 'hot spots' of activity. Hot spot analyses are particularly useful for urban planning applications because they allow for detection of subregions, or clusters, and can provide the location and extent of aggregations of extreme values or give an indication of the heterogeneity existing across the area (Anselin 1995).

While both global and local autocorrelation measures are used frequently in shrinking urban applications (Riguelle, Thomas, and Verhetsel 2007; Frazier, Bagchi-Sen, and Knight 2013; Murgante and Rotondo 2013), the benefit of local clustering tests over global statistics is that they may identify significant, but smaller, pockets of local clustering within the city – the intra-region variability discussed above. Global statistics are limited because they cannot provide information regarding the size or location of specific pockets of raised incidence (Rogerson and Yamada 2009). For example, the Gi* statistic will produce high values along with a high positive z-score when there is a dominant pattern of high values near other high values and will produce low values when there is clustering of low values (Rogerson and Yamada 2009). Furthermore, tests such as the Gi* statistic can identify both "hot" and "cold" spots – or areas of high and low autocorrelation. The use of local clustering tests can eliminate biases that can arise when specific areas are selected for testing based on preconceptions about their existence (Ord and Getis 1995).

Global and local forms of spatial autocorrelation are often used together to develop a comprehensive picture of urban shrinkage and decline across the city at various scales. For example, in their analysis of the deindustrialized city of Taranto in southern Italy, Murgante and Rotondo (2013) used the global Moran's I statistic as well as its local form, LISA, in order to determine whether there was spatial clustering across the city for six indicators of shrinkage: dependency ratio (i.e., ratio of dependents to independents), unemployment rate, renter occupation, foreign population, educational attainment, and persons per household. The authors found positive global spatial autocorrelation for all six variables across this shrinking city, meaning high unemployment rates were located near to other high unemployment rates, and so on. Furthermore, using the local LISA statistic, the authors were able to identify specific pockets within the city that were likely experiencing "hot spots" of shrinkage – or extremely raised incidences – due to the local clustering of high levels of these six indicators.

Similarly, both local and global clustering techniques were used by Frazier, Bagchi-Sen, and Knight (2013) to analyze the social and environmental impacts of building demolition on crime in Buffalo, New York. Using the global form of the Getis-Ord statistics, the authors identified a threshold for observing spatial associations of demolition and crime within the city and then used those distances to identify local hot spots of demolition and crime activity throughout the city. Comparing the hot spots of demolition to those of crime across a five-year period, the authors found that as demolition activities were targeted to a certain area of the city, which is often the case in shrinking cities, certain crimes migrated away from that area. In particular, drug arrests and prostitution, two types of crime that are known for utilizing vacant and abandoned structures, showed significant declines in the areas where demolitions were targeted. Their study showed through geospatial clustering techniques how the process of demolition can not only change the physical structure and pattern of the urban fabric but may also impact the localized social processes taking place in certain neighborhoods. Furthermore, the authors noted that as demolition activities progressed across a five-year time period, the mean center of criminal activity moved toward the periphery of the city limits (and very likely moved into the first-ring suburbs). The spatial clustering techniques were able to demonstrate empirically the "spillover" effects of shrinkage and decline-related issues into surrounding municipalities, reinforcing the need for strong governance structures discussed in Chapter 8. Furthermore, because the study was undertaken at a sub-city scale (the geographical unit of analysis was the block group), certain implications of "rightsizing" (Ch. 6) through demolition were highlighted that would not have been possible through a global statistic.

Modifiable areal unit problem (MAUP) and other considerations

Moving forward, scholarly activity focused on shrinking and declining places would benefit from more localized spatial analysis of the emerging variables discussed above. However, there are several limitations of these types of spatial statistics that deserve attention. In particular, whenever spatial data are aggregated into areal units, and those aggregate units are used for analysis, statistical biases may affect any subsequent analyses. These biases form the basis of the theoretical foundation known as the modifiable areal unit problem (MAUP; Openshaw and Taylor 1979; Openshaw 1984). There are two main issues associated with MAUP. The first is the effect of scale, which results from the selected areal unit at which mapping or analysis is completed. Whenever point data are aggregated to areal units such as census block groups, the locations of those point data are no longer relevant, and the areal unit is considered "homogenous". This type of aggregation introduces biases into spatial analyses performed using the data since the original locations of the data are no longer considered. The second MAUP issue is a grouping or zonation problem, which results when a larger number of smaller units are grouped or aggregated into a smaller number of larger units (Openshaw and Taylor 1979). The method of aggregation that is used to compile the data (e.g., mean, median, majority rules, etc.) can also introduce statistical biases into the analysis. These MAUP biases are also familiar to many scientists as ecological fallacies, and while investigators have been working to overcome them for decades (see Openshaw et al. 1987; Besag and Newell 1991; Gatrell et al. 1996), no universal solution has been reached.

Additionally, many cluster detection techniques rely on adjacent neighboring areas to provide context. When the geographic area of analysis is restricted to a bounded window, such as city limits, several different types of *edge effects* can occur. First, there can be sampling bias, which occurs when the probability of observing an object depends on its shape or size. Rogerson and Yamada (2009) found that these biases can typically be remedied by weighting observations accordingly. The second type of edge effects are censoring effects, which occur when the extent of an object or phenomena that lies partially within the window cannot be observed in full. For example, a cultural neighborhood may straddle two different administrative districts, and limiting the analysis to a single district or splitting the social variable across both districts will lead to censoring effects where the entire neighborhood is not being analyzed appropriately. Censoring effects can be remedied by delineating "guard" or "buffer" areas, which are used for boundary estimation but are not presented in results or used in subsequent analysis (Vidal Rodeiro and Lawson 2005; Van Meter et al. 2010). The use of guard areas is particularly useful in areal and cluster analyses (discussed below), but it can be difficult to obtain compatible data across municipal boundaries. In fact, a major issue in shrinking cities is the fragmentation of governance (Ch. 8), which unfortunately often also accompanies a fragmentation in available data.

Final remarks

In this chapter, we sought to summarize the major points of the book and shed light on emerging future areas of research for shrinking cities. The take-home message is that, moving forward, we may need to look *within* the demographics and economics of cities – through both qualitative and empirical approaches – to truly understand the complex underlying processes driving shrinkage and decline. The demographics of a place are rarely homogenous, and similarly, the demographics between two cities are almost never the same. By seeking to understand the within-group population variation at finer scales within the city, we may start to understand the unique processes driving population loss, economic gains, or sustainability and resilience at the ground level. However, we need not stop with basic breakdowns of race or ethnicity. Within these groups we will find variations that may help explain why a certain place remained stable while another struggled. With the theoretical foundation discussed in this book, along with a toolbox of methods for empirical analysis, the challenge is to incorporate these emerging debates into future shrinking cities research.

References

Airriess, Christopher A. 2016. *Contemporary Ethnic Geographies in America*. 2nd ed. New York: Rowman & Littlefield.

Anselin, Luc. 1995. "Local Indicators of Spatial Association-Lisa." *Geographical Analysis* 27 (2):93–115.

Anselin, Luc. 1996. "The Moran Scatterplot as an ESDA Tool to Assess Local Instability in Spatial Association." In *Spatial Analytical Perspectives on GIS in Environmental and Socio-Economic Sciences*, edited by Fischer, M., Scholten, H., and Unwin, D., 111–125. London: Taylor and Francis.

Beauregard, Robert A. 2009. "Urban Population Loss in Historical Perspective: United States, 1820–2000." *Environment and Planning A* 41 (3):514–528.

Berube, Alan. 2014. *All Cities Are Not Created Unequal*. Washington, DC: Brookings Institution.

Besag, Julian, and James Newell. 1991. "The Detection of Clusters in Rare Diseases." *Journal of the Royal Statistical Society. Series A (Statistics in Society)* 154 (1):143–155.

Chetty, Raj, Nathaniel Hendren, and Lawrence F. Katz. 2015. "The Effects of Exposure to Better Neighborhoods on Children: New Evidence from the Moving to Opportunity Experiment." Accessed Jan 26. http://www.equality-of-opportunity.org/images/mto_paper.pdf.

Chetty, Raj, Nathaniel Hendren, Patrick Kline, and Emmanuel Saez. 2014. "Where Is the Land of Opportunity? The Geography of Intergenerational Mobility in the United States." Accessed Jan 26. http://www.equality-of-opportunity.org/images/mobility_geo.pdf.

Ehrenhalt, Alan. 2012. *The Great Inversion and the Future of the American City*. New York: Vintage Books.

Frazier, Amy E., Sharmistha Bagchi-Sen, and Jason Knight. 2013. "The Spatio-Temporal Impacts of Demolition Land Use Policy and Crime in a Shrinking City." *Applied Geography* 41:55–64.

Gatrell, Anthony C., Trevor C. Bailey, Peter J. Diggle, and Barry S. Rowlingson. 1996. "Spatial Point Pattern Analysis and Its Application in Geographical Epidemiology." *Transactions of the Institute of British Geographers* 21 (1):256–274.

Getis, Arthur, and J. Keith Ord. 1992. "The Analysis of Spatial Association by Use of Distance Statistics." *Geographical Analysis* 24 (3):189–206.

Glaeser, Edward. 2011. *Triumph of the City: How Our Greatest Invention Makes Us Richer, Smarter, Greener, Healthier, and Happier*. London, UK: Penguin Books.

Godfrey, Brian. 2016. "New Ethnic Landscapes: Place Making in Urban America." In *Contemporary Ethnic Geographies in America*, edited by Christopher A. Airriess, 59–90. New York: Rowman & Littlefield.

Johnson, Kenneth M. 2013a. *Deaths Exceed Births in Record Number of U.S. Counties*. Durham, NH: Carey Institute, University of New Hampshire.

Johnson, Kenneth M. 2013b. "Demographic Trends in Nonmetropolitan America: Implications for Land Use Development and Conservation." Sociology Scholarship, Paper 46. http://scholars.unh.edu/soc_facpub/46.

Mallach, Alan. 2015. "The Uncoupling of the Economic City: Increasing Spatial and Economic Polarization in American Older Industrial Cities." *Urban Affairs Review* 51 (4):443–473.

Moran, Patrick A.P. 1950. "Notes on Continuous Stochastic Phenomena." *Biometrika* 37 (1):17–23.

Murgante, Beniamino, and Francesco Rotondo. 2013. "A Geostatistical Approach to Measure Shrinking Cities: The Case of Taranto." In *Statistical Methods for Spatial Planning and Monitoring*, edited by Silvestro Montrone and Paola Perchinunno, 119–142. Milan: Springer Milan.

Openshaw, Stan. 1984. "Ecological Fallacies and the Analysis of Areal Census Data." *Environment and Planning A* 16 (1):17–31.

Openshaw, Stan, Martin Charlton, Colin Wymer, and Alan Craft. 1987. "A Mark 1 Geographical Analysis Machine for the Automated Analysis of Point Data Sets." *International Journal of Geographical Information Systems* 1 (4):335–358.

Openshaw, Stan, and Peter J. Taylor. 1979. "A Million or So Correlation Coefficients: Three Experiments on the Modifiable Areal Unit Problem." *Statistical Applications in the Spatial Sciences* 21:127–144.

Ord, J. Keith, and Arthur Getis. 1995. "Local Spatial Autocorrelation Statistics: Distributional Issues and an Application." *Geographical Analysis* 27 (4):286–306.

Riguelle, François, Isabelle Thomas, and Ann Verhetsel. 2007. "Measuring Urban Polycentrism: A European Case Study and Its Implications." *Journal of Economic Geography* 7 2):193–215

Rogerson, Peter, and Ikuho Yamada. 2009. *Statistical Detection and Surveillance of Geographic Clusters*. Boca Raton, FL: CRC Press.

Rogerson, Peter A. 2015. *Statistical Methods for Geography: A Student's Guide*. Washington, DC: Sage Publications.

Sawhill, Isabel. 2015. "Inequality and Social Mobility: Be Afraid." Brookings Institution. Accessed Jan 26. http://www.brookings.edu/blogs/social-mobility-memos/posts/2015/05/27-inequality-great-gatsby-curve-sawhill.

Skop, Emily. 2016. "Asian Indians: Construction of Community and Identity." In *Contemporary Ethnic Geographies in America*, edited by Christopher A. Airriess, 325–346. New York: Rowman & Littlefield.

Tippett, Rebecca. 2015a. "2010–2014 County Population Change and Components of Change. Carolina Demography." Carolina Population Center, UNC-Chapel Hill. Accessed Jan 26. http://demography.cpc.unc.edu/2015/04/13/2010–2014-county-population-change-and-components-of-change/.

Tippett, Rebecca. 2015b. "Nationwide, Majority of Counties Have Lost Population since 2010. Carolina Demography." Carolina Population Center, UNC-Chapel Hill. Accessed Jan 26. http://demography.cpc.unc.edu/2015/04/06/nationwide-majority-of-counties-have-lost-population-since-2010/.

Van Meter, Emily M., Andrew B. Lawson, Natalie Colabianchi, Michele Nichols, James Hibbert, Dwayne E. Porter, and Angela D. Liese. 2010. "An Evaluation of Edge Effects in Nutritional Accessibility and Availability Measures: A Simulation Study." *International Journal of Health Geographics* 9 (40):1–12.

Vidal Rodeiro, Carmen L., and Andrew B. Lawson. 2005. "An Evaluation of the Edge Effects in Disease Map Modelling." *Computational Statistics & Data Analysis* 49 (1):45–62.

Wilson, Jill H., and Nicole Prchal Svajlenka. 2014. "Immigrants Continue to Disperse, with Fastest Growth in the Suburbs." In *Immigration Facts Series*. Washington, DC.

Appendix A

As mentioned in Chapter 3, a *geometric mean* is an average obtained through multiplication rather than addition. Such a mean is useful when the objective is to account for any compounding effects that might exist when all the constituent parts of the mean are found together (Spizman and Weinstein 2008). Recall that the G-index developed in Chapter 3 seeks to measure *concentrated disadvantage* (CD). One reason for an interest in CD in urban studies is that, in neighborhoods characterized by multiple forms of deprivation or disadvantage, "difficulties reinforce one another to produce a situation of compound disadvantage" (Pacione 2003: 326–327). It follows that an index of CD which attempts to account for this compounding effect is likely to be a more comprehensive indicator of veritable *distress* than simple poverty or unemployment rates alone.

That being said, geometric means are subject to a "zero problem" – because they are computed by way of multiplication, a single zero value in one part of the formula results in an overall geometric mean of zero, even if the values for the other parts (variables) in the formula are all meaningfully greater than zero. Recognizing that this property of geometric means poses a problem for computing CD in, say, racially homogenous white census tracts (see Chapter 3), we use the multiplicative zero replacement strategy for computing geometric means proposed by Zeng and colleagues (2014). Specifically, each of the five variables (x) that were used to calculate our G-index of concentrated disadvantage were first transformed as follows: Zeng and colleagues (2014)

$$x'_i = \begin{cases} \delta, x_i = 0 \\ x_i \left(1 - \sum_{x_j=0} \delta\right), x_i \neq 0 \end{cases}$$

where x' is the transformed value of the original variable, x (e.g., the percent of the population that is non-white), i is an index over n census tracts, and δ is a replacement value, which in this case is set to 0.0001.

Adopting this zero replacement strategy allows us to consistently compute non-zero G-indices for all census tracts in our study. For this reason, however, the lower limit on the G-index will only approach zero, not equal zero.

Appendix B

In general, a McNemar test is a chi-squared test that is applied to contingency tables in which the row and column data are dependent rather than independent. That is, in conventional chi-squared tests for independence, such as the one carried out for the relationship between population and economic shrinkage in Table 2.6, row and column data are assumed to be independent (the null hypothesis). Rejecting the null hypothesis therefore allows a researcher to conclude, as we did for Table 2.6, that two variables are dependent or move together. By contrast, in a McNemar test, the row and column data are assumed to be correlated or paired from the outset. For instance, a census tract might be classified as *shrinking* based on its four-decade (1970–2010) annual rate of exponential population change; and it might or might not be classified as *on pace to be shrinking* on the basis of its annual rate of exponential population change over the most recent decade (2000–2010). Put differently, a given tract's status as *shrinking* can be different depending on whether the period of population change analysis begins in 1970 or in 2000. A McNemar test allows researchers to evaluate whether, in the aggregate, tracts' *shrinking* statuses did in fact change from 1970 to 2000. The row variable in such a test is the tract-level shrinkage classification based on a starting period of 1970, and the column variable is the analogous classification for a starting period of 2000. Hence, we are dealing with the same sample of tracts but observing its shrinkage status at two points in time (i.e., the data are dependent).

Rather than illustrating a one-sample McNemar test of the type described above, however, we move directly to the type of two-sample McNemar test that is carried out. In brief, a two-sample extension of the McNemar test tests the null hypothesis that patterns of change (see the preceding paragraph) are the same across two groups. The two-sample test therefore begins by conducting separate McNemar tests for each group involved in the analysis, in order to identify "discordant pairs", or tracts whose shrinkage statuses changed over time (Adedokun and Burgess 2011). Tables B.1 through B.3 illustrate how these tests are set up for the analyses that are described in Chapter 5. First, the sample of census tracts is split into two groups: tracts within core cities, and tracts that are outside core cities. Second, for each group, tract classifications are summarized in contingency tables according to their time-varying statuses as shrinking (Table B.1), declining (Table B.2), and both shrinking and declining (Table B.3). These group-specific

Table B.1 Two-sample McNemar test for change in tract-level patterns of shrinkage

	In Central City		*Not In Central City*	
	Shrank (2000–2010)	*Did Not Shrink (2000–2010)*	*Shrank (2000–2010)*	*Did Not Shrink (2000–2010)*
Shrank (1970–2010)	2,928	**2,208**	936	**1,314**
Did Not Shrink (1970–2010)	**2,975**	16,183	**2,088**	23,747

Notes: McNemar tests for both contingency tables are significant with $p << 0.001$, suggesting that overall patterns of population shrinkage have changed both inside and outside central cities; **bold text** indicates discordant pairs (i.e., tracts whose shrinkage statuses for the period 1970–2010 were different from their corresponding statuses for the more recent period 2000–2010)

Table B.2 Two-sample McNemar test for change in tract-level patterns of decline

	In Central City		*Not In Central City*	
	Declined (2000–2010)	*Did Not Decline (2000–2010)*	*Declined (2000–2010)*	*Did Not Decline (2000–2010)*
Declined (1970–2010)	2,209	**3,444**	1,073	**1,406**
Did Not Decline (1970–2010)	**1,166**	17,164	**1,523**	23,787

Notes: McNemar tests for both contingency tables are significant with $p << 0.001$, suggesting that overall patterns of decline have changed both inside and outside central cities; **bold text** indicates discordant pairs (i.e., tracts whose decline statuses for the period 1970–2010 were different from their corresponding statuses for the more recent period 2000–2010)

Table B.3 Two-sample McNemar test for change in tract-level patterns of coupled shrinkage and decline

	In Central City		*Not In Central City*	
	Shrank-Declined (2000–2010)	*All Other Tracts (2000–2010)*	*Shrank-Declined (2000–2010)*	*All Other Tracts (2000–2010)*
Shrank-Declined (1970–2010)	496	**1,054**	95	**207**
All Other Tracts (1970–2010)	**510**	21,926	**295**	27,192

Notes: McNemar tests for both contingency tables are significant with $p << 0.001$, suggesting that overall patterns of coupled shrinkage and decline have changed both inside and outside central cities; **bold text** indicates discordant pairs (i.e., tracts whose shrinkage-decline statuses for the period 1970–2010 were different from their corresponding statuses for the more recent period 2000–2010) (dependent data) contingency tables allow us to identify the overall number of tracts whose classifications did in fact change over time. These "discordant pairs" are flagged with bold text in each of the two-sample McNemar tables reported below. These bold numbers (discordant pairs), in turn, are the figures found in Tables 5.4–5.6. The discordant pairs allow us to understand whether patterns of status change are the same inside and outside central cities. As we demonstrate in Chapter 5, tracts outside central cities are tending toward shrinkage, decline, and coupled shrinkage and decline much more significantly than tracts within core cities.

Appendix C

Income inequality in America's 50 largest cities, 2007–2012

	City	*Population*	*Household Income, 2012*		*Ratio 2012*	*Change in Household Income, 2007–2012*		*Ratio Change, 2007–2012*	
			20th percentile	*95th percentile*		*20th percentile*	*95th percentile*		
1	Atlanta, Georgia	443,768	$14,850	$279,827	18.8	–$4,036	–$16,813	3.1	
2	San Francisco, California	825,863	$21,313	$353,576	16.6	–$4,309	$27,815	3.9	*
3	Miami, Florida	413,864	$10,438	$164,013	15.7	–$1,840	–$3,397	2.1	*
4	Boston, Massachusetts	637,516	$14,604	$223,838	15.3	–$1,359	–$14,912	0.4	
5	Washington, D.C.	632,323	$21,782	$290,637	13.3	–$22	$7,645	0.4	
6	New York, New York	8,336,697	$17,119	$226,675	13.2	–$1,735	–$8,677	0.8	
7	Oakland, California	400,740	$17,646	$223,965	12.7	–$1,062	–$14,059	0.0	
8	Chicago, Illinois	2,714,844	$16,078	$201,460	12.5	–$2,194	–$4,100	1.3	*
9	Los Angeles, California	3,857,786	$17,657	$217,770	12.3	–$3,107	–$26,242	0.6	
10	Baltimore, Maryland	621,342	$13,522	$164,995	12.2	–$2,706	–$7,586	1.6	*

(*Continued*)

(Continued)

	City	*Population*	*Household Income, 2012*		*Ratio 2012*	*Change in Household Income, 2007–2012*		*Ratio Change, 2007–2012*	
			20th percentile	*95th percentile*		*20th percentile*	*95th percentile*		
11	Houston, Texas	2,161,686	$17,344	$205,490	11.8	–$1,977	–$10,327	0.7	
12	Philadelphia, Pennsylvania	1,547,607	$12,850	$151,026	11.8	–$1,536	$2,638	1.4	*
13	Dallas, Texas	1,241,108	$17,811	$200,367	11.2	–$2,392	–$25,065	0.1	
14	Detroit, Michigan	701,524	$9,083	$101,620	11.2	–$2,098	–$19,820	0.3	
15	Minneapolis, Minnesota	392,871	$17,753	$193,777	10.9	–$1,486	–$8,256	0.4	
16	Memphis, Tennessee	655,141	$13,520	$145,015	10.7	–$1,231	–$12,014	0.1	
17	Cleveland, Ohio	390,923	$9,432	$100,903	10.7	–$1,865	–$5,537	1.3	*
18	Tulsa, Oklahoma	394,098	$17,359	$183,407	10.6	$38	$4,127	0.2	
19	Denver, Colorado	634,265	$19,770	$208,810	10.6	$1,000	$7,169	–0.2	
20	Fresno, California	505,870	$15,665	$160,360	10.2	–$3,257	–$6,171	1.4	*
21	Charlotte, North Carolina	775,208	$21,998	$219,126	10.0	–$4,864	–$6,815	1.6	*
22	Kansas City, Missouri	464,346	$16,353	$161,488	9.9	–$1,641	–$2,668	0.8	
23	Long Beach, California	467,888	$19,255	$185,543	9.6	–$3,042	–$14,302	0.7	
24	Austin, Texas	842,595	$21,738	$207,594	9.5	–$1,646	–$10,787	0.2	
25	Portland, Oregon	603,650	$20,152	$191,492	9.5	–$1,535	$3,681	0.8	
26	Tucson, Arizona	524,278	$13,798	$130,327	9.4	–$3,800	–$9,029	1.5	*
27	Sacramento, California	475,524	$17,901	$168,858	9.4	–$6,608	–$12,393	2.0	*
28	Milwaukee, Wisconsin	598,920	$13,328	$125,363	9.4	–$3,481	$237	2.0	*
29	El Paso, Texas	672,534	$16,206	$151,745	9.4	$1,530	–$4,486	–1.3	
30	Indianapolis, Indiana	835,806	$16,230	$150,346	9.3	–$5,811	–$16,883	1.7	*

31	Seattle, Washington	634,541	$26,156	$239,549	9.2	–$678	–$11,471	–0.2	
32	Louisville, Kentucky	605,108	$16,924	$152,792	9.0	–$1,636	–$11,832	0.2	
33	Albuquerque, New Mexico	555,419	$18,646	$168,121	9.0	–$2,818	–$239	1.2	*
34	Nashville, Tennessee	623,255	$18,539	$166,032	9.0	–$3,914	–$10,293	1.1	*
35	San Diego, California	1,338,354	$25,126	$224,814	8.9	–$3,158	–$13,942	0.5	
36	San Jose, California	982,783	$31,047	$273,766	8.8	–$3,560	$8,143	1.1	
37	Jacksonville, Florida	836,507	$17,411	$152,329	8.7	–$7,843	–$18,999	2.0	*
38	Phoenix, Arizona	1,488,759	$19,186	$167,503	8.7	–$3,796	–$26,099	0.3	
39	San Antonio, Texas	1,383,194	$18,518	$158,566	8.6	–$1,480	–$5,381	0.4	
40	Columbus, Ohio	809,890	$17,238	$147,496	8.6	–$1,134	$1,295	0.6	
41	Oklahoma City, Oklahoma	599,309	$18,835	$160,125	8.5	–$1,492	–$12,331	0.0	
42	Raleigh, North Carolina	423,743	$24,113	$199,911	8.3	–$1,137	–$174	0.4	
43	Omaha, Nebraska	421,564	$19,649	$161,910	8.2	–$2,252	–$7,658	0.5	
44	Fort Worth, Texas	782,027	$20,992	$168,989	8.1	–$1,701	–$827	0.6	
45	Colorado Springs, Colorado	431,846	$22,213	$175,034	7.9	–$3,372	–$4,378	0.9	*
46	Wichita, Kansas	385,586	$19,516	$151,068	7.7	–$2,781	–$16,879	0.2	
47	Las Vegas, Nevada	596,440	$21,380	$164,344	7.7	–$6,248	–$36,330	0.4	
48	Mesa, Arizona	452,068	$21,007	$157,190	7.5	–$5,952	–$10,044	1.3	*
49	Arlington, Texas	375,598	$24,169	$175,759	7.3	–$3,458	$220	0.9	*
50	Virginia Beach, Virginia	447,021	$31,051	$187,652	6.0	–$4,727	$211	0.8	*

Data source: Brookings institution analysis of 2007 and 2012 American Community Survey data

* Change is statistically significant at the 95% confidence interval

Index